CRUELTY

CRUELTY

HUMAN EVIL AND THE HUMAN BRAIN

KATHLEEN TAYLOR

OXFORD
UNIVERSITY PRESS

OXFORD

UNIVERSITY PRESS

Great Clarendon Street, Oxford OX2 6DP

Oxford University Press is a department of the University of Oxford.
It furthers the University's objective of excellence in research, scholarship,
and education by publishing worldwide in

Oxford New York

Auckland Cape Town Dar es Salaam Hong Kong Karachi
Kuala Lumpur Madrid Melbourne Mexico City Nairobi
New Delhi Shanghai Taipei Toronto

With offices in

Argentina Austria Brazil Chile Czech Republic France Greece
Guatemala Hungary Italy Japan Poland Portugal Singapore
South Korea Switzerland Thailand Turkey Ukraine Vietnam

Oxford is a registered trademark of Oxford University Press
in the UK and in certain other countries

Published in the United States
by Oxford University Press Inc., New York

© Kathleen Taylor 2009

The moral rights of the author have been asserted
Database right Oxford University Press (maker)

First published 2009

British Library Cataloguing in Publication Data

Data available

Library of Congress Cataloging in Publication Data

Data available

Typeset by SPI Publisher Services, Pondicherry, India
Printed in Great Britain
on acid-free paper by
CPI Antony Rowe, Chippenham, Wiltshire

ISBN 978–0–19–955262–7

1 3 5 7 9 10 8 6 4 2

For my grandmother
who, like many before and since, asked:
'How can people do such awful things?'

Preface and Acknowledgements

Cruelty is a book on a difficult topic. Myths and stereotypes abound, opinions are often strongly held, and the relevant literature is staggeringly vast. In writing such a book any author is indebted to innumerable others. Memory is fallible, and ideas can sometimes resurface in the mind having lost the label which names their originator. If any reader feels that I have failed to acknowledge my predecessors in full, they are thus almost certainly right. I have nonetheless provided extensive referencing in the notes, so that readers can track and check sources should they wish to do so.

Likewise, it has not been possible to reference every major atrocity, even among those committed in recent years. Rare though mass cruelty may be, its manifestations are still too numerous to list exhaustively here. My omission of, for example, the approximately 30,000 'disappeared' (*los desaparecidos*) of Argentina reflects no downgrading of that particular tragedy—this author is wary of attempts to rank atrocities. Nor is any slight intended to the victims, living or dead, of the horrific incidents omitted from these pages.

Cruelty's intended audience does not necessarily work in academia, have scientific training, or have a professional interest in the topic. Accordingly, I have tried to keep the book readable, confining jargon to the notes where possible and explaining it where necessary. That is not to say, however, that the content is easy; it isn't. Human cruelty lies at the interface between morality and science, and the ideas touched on here go very deep. I hope that readers will find the journey exciting rather than overwhelming.

There is also, of course, the viscerally disturbing nature of the material to consider. Examples are needed to remind us why cruelty matters, but I have thought long and hard about including each one. They are not there to display the kind of theoretical bravado which says 'Look at how much horror I can handle', but to anchor theoretical discussion to its brutal and sickening reality. *Cruelty* is not unremittingly bleak (careful readers may even find the occasional small joke), but—well, the clue is in the title.

Among the many who have assisted me, I would particularly like to thank George Kassimeris (University of Wolverhampton), Richard Overy (University of Exeter), Daniel Statman (University of Haifa), and the anonymous readers of the book at its early stages for their extensive support and assistance in clarifying my ideas. The Oxford University Press Delegates, who approved the book for publication, also provided useful suggestions. In Oxford, John Stein and Peter Hansen took the time to engage in fascinating discussions, as did Harvey Whitehouse, Miguel Farias, Rebecca Roache, Guy Kahane, Nick Shackel, and Katja Wiech. I am also grateful to the staff of Oxford's libraries—especially those of the wonderful Bodleian and Taylorian libraries—who helped me track down references. Carolyn Korsmeyer (University at Buffalo) helped with Lacan, Seth Maislin (Potomac Indexing, LLC) provided advice about indexing, Randall Bykwert (Calvin College) allowed me to use his image from *Der Stürmer*, and Susan Greenfield, Bernard Gesch, and Eva Cylharova (Oxford) offered welcome moral support, as did Alison and David Taylor. Gillian Wright gave me invaluable help and support throughout, not least through her scrupulous proof-reading. I am also grateful to my agent, Catherine Clarke, and to others at Felicity Bryan.

Finally, I owe a great deal to the excellent team at Oxford University Press (James Thompson, Kate Farquhar-Thompson, Phil Henderson, Jeff New, and others), and especially to *Cruelty*'s editor Latha Menon, whose guidance, encouragement, and enthusiasm for science have been welcome sustenance during this challenging project.

Contents

CONTENTS

X

List of figures

Introduction

Cruelty in context

I'the last night's storm I such a fellow saw,
Which made me think a man a worm.
[...]
As flies to wanton boys are we to the gods,
They kill us for their sport.

(William Shakespeare, King Lear)

The fearful asymmetry of birth and death

Human beings are difficult creatures to build. Until the recent promise of reproductive technology emerged to tantalize, and sometimes fulfil, the unhappily childless, new members joined the *Homo sapiens* club through only one mechanism: sex. And membership was far from guaranteed. Around three-quarters of women who conceive lose the pregnancy before they realize they are pregnant. Around a sixth of recognized pregnancies fail.[1] God may disapprove of abortion, as some of his followers claim, but nature seems to use it frequently. Even in the womb, that symbol of safety, we are fragile creatures, difficult to sustain, easily nudged across the border between life and death.

When it comes to leaving Club Human the contrast is clear: one entrance, a myriad exits.[2] Complex entities have more opportunities to malfunction, and as for your computer, so for you. Even among those humans who make it to babyhood, let alone adulthood, normality is hard to define and perfection impossible to achieve. All of us have our blemishes, some physical, some psychological, some visible, some less so. Among these knots in the weave are deadly faults. Christianity speaks of original sin and the resulting corruption, passed down from generation to generation, which brings ageing, pain, and death to humankind. Modern science names our genes as the secular equivalent—bearers of an intricate and individual recipe for the bodily dysfunction which consumes us sooner or later, depending on how stressed we are, whether we smoke or drink alcohol, what we eat and how much exercise we take. Obvious physical problems with internal organs—heart or liver failure, cancers and immune disorders, genetic conditions, strokes or seizures, neurodegenerative disorders and blood clots—account for many of us. Infectious diseases remove many more. Then there are the less well understood but often lethal disorders we shove into the pigeonhole of 'mental illness': depression, schizophrenia, anorexia, and suchlike. Our bodies may grow according to a standard template, but they can fail in an oppressive variety of ways. Human existence is framed by this fearful asymmetry between creation and destruction.

An atheist might be tempted—and many have been—to sneer at the notion of a loving God who provides so many awful ways to die. Mockers should pause, however; because the inventiveness of nature as thanatologist is rivalled by the inventiveness of man.[3] I use the older, generic term with the obvious implication: men are far more likely to wreak violent death on others than are women. That is not to say that women cannot fight wars, commit murders, or dream up hideous tortures; they can and do. But most lethal violence comes from men.[4]

And that violence can be rampantly baroque. Humans are staggeringly creative when it comes to hurting and killing other humans. You don't need to read the Marquis de Sade to know this; you only need to listen to the news, read fiction, watch movies or TV, or learn a little history. A news report from Lebanon describes a body retrieved from under a bombed-out building as 'paper-thin'; presumably *that particular demise* was not wished on *that particular individual* by the pilots who dropped the annihilating bombs, but the poor man is no less dead for that, his body no less horrifically desecrated.[5] A novel, soon filmed,

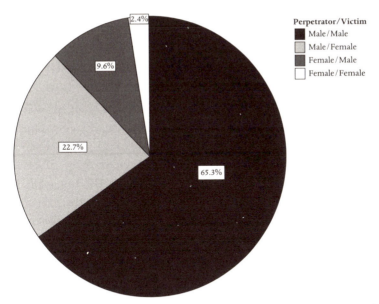

Perpetrator/Victim
- Male/Male
- Male/Female
- Female/Male
- Female/Female

FIGURE I. The gender balance for homicide, the archetypal violent crime, as recorded by the United States government for the year 2005. The pie chart shows the proportions of homicides committed by men against men (black segment, 65.3%), men against women (light grey segment, 22.7%), women against men (Medium grey segment, 9.6%), and women against women (white segment, 2.4%). In other words, men were the perpetrators in just under nine-tenths (88%) of the crimes recorded, and about three quarters of their victims (74.2%) were men. Only 12% of homicides had female perpetrators; four-fifths of their victims (80%) were men.

These proportions for 2005 are comparable to the US government's summary statistics for the last three decades (1976–2005), which found that men perpetrated 88.8% of all homicides, and that 76.5% of all homicide victims were men. (Further information on the data gathering and analysis techniques used can be found at the the web site http://www.ojp.usdoj.gov/bjs/homicide/homtrnd.htm#contents.)

describes a serial killer who de-skulls a living victim, scooping out a spoonful of brain to cook.[6] As for the supposedly less sensational pages of history's archives, they are drenched in blood, bursting with examples of hideous suffering: children mutilated and dismembered; adults flayed or disembowelled; women raped to death; the severed heads of suicide bombers found among their victims' fragments; foetuses cut from their murdered mothers' wombs; the vivisection or intentional infection of prisoners; people burned or boiled or buried or shredded alive, ripped apart by animals or machines, strangled or drowned or suffocated, nailed or tied or hung up and left to die, disintegrated by an atomic blast or blown to bits by more conventional explosives.

The problem of human evil

Religious believers in a benevolent deity presumably accept that the goodness in people—love, kindness, generosity, and so on—is sufficient to outweigh, in whatever manner God makes such assessments, the evil that they do to one another. Somewhere the books must balance. Looking at just how much cruelty we come up with, it's hard to see where (unless there are some truly delightful aliens out there, in which case let's hope for their sake they never meet us). You may, of course, put your trust in the post-mortem justice of heaven and hell, but these are problematic comforts. In particular, hell—which entails inflicting yet more cruelty, *for ever*, on those who were cruel (or adulterous or sodomites or whatever) during their relatively brief lives—raises more questions than it answers. Why would a loving and omnipotent divinity not simply prevent people from being cruel in the first place?

We have reached well-trodden ground, for this is the problem of evil which has bedevilled religion—particularly the great monotheisms, which insist on God's omnipotence and goodness—for centuries. It has driven some believers out of the fold, leading them to conclude that the human race is, all told, a really rather vile and disgusting species, inherently bad, fit only for whips and chains. Others resort to the language of mystery, citing God's inscrutability, or claims about the necessity of suffering for spiritual growth. Plausible enough in the cool air of academe, these tend to burn up in the heat of atrocity.[7]

One of the Marquis de Sade's female villains, the murderous Mme Dubois of his novel *Justine*, formulates the problem of evil bluntly:

> 'I believe,' this dangerous woman answered, 'that if there were a God there would be less evil on earth; I believe that since evil exists, these disorders are either expressly ordained by this God, and there you have a barbarous fellow, or he is incapable of preventing them and right away you have a feeble God; in either case, an abominable being, a being whose lightning I should defy and whose laws

contemn . . . is not atheism preferable to the one and the other of these extremes?'[8]

Cruelty is the epitome of human evil.[9] We may condemn neglect or theft, express our fury at liars and cheats, or shudder at other people's sexual practices, but we are probably aware that what we call malpractice or perversion may not be so defined by everyone. Cruelty, however, carries a moral weight which makes it hard to resist. People all over the world react with horror, rage, pity, grief, and disgust to stories of atrocity. Although, as we shall see, that horror may be distorted or suppressed under certain conditions, it is a natural and powerful response. The moral judgements it provokes seem to rely on similar criteria in different cultures. Innocent and defenceless victims, gratuitous torture, enjoyment on the part of the perpetrator—these are the cues which provoke the harshest condemnation of cruelty.

When we think of cruelty we often think of its most virulent extremes: the bizarre sadism of a serial killer or the mass slaughter of war. A representative example, which readily conjures the epithet of 'evil', comes from one of the twentieth century's less well known human-rights disasters: the three-and-a-half-decade-long internal conflict in Guatemala. Over 42,000 people are thought to have been killed. Yes, you may find yourself thinking, people die all the time, especially in civil wars and insurgencies—especially, perhaps, in nations not quite as civilized as ours. True. But in acknowledging the numbers it can be easy to lose sight of *how* those unfortunate innocents lost their lives:

> The Army's perception of Mayan communities as natural allies of the guerrillas contributed to increasing and aggravating the human rights violations perpetrated against them, demonstrating an aggressive racist component of extreme cruelty that led to the extermination en masse, of defenceless Mayan communities purportedly linked to the guerrillas—including children, women and the elderly—through methods whose cruelty has outraged the moral conscience of the civilised world. . . . In the majority of massacres there is evidence of multiple acts of savagery, which preceded, accompanied or occurred after the deaths of the victims. Acts such as the killing of defenceless children, often by beating them against walls or throwing them alive

into pits where the corpses of adults were later thrown; the amputation of limbs; the impaling of victims; the killing of persons by covering them in petrol and burning them alive; the extraction, in the presence of others, of the viscera of victims who were still alive; the confinement of people who had been mortally tortured, in agony for days; the opening of the wombs of pregnant women, and other similarly atrocious acts, were not only actions of extreme cruelty against the victims, but also morally degraded the perpetrators and those who inspired, ordered or tolerated these actions.[10]

Most of us find it difficult or impossible to imagine what it would feel like to do such things to living human beings, just as we find it hard to understand sadistic serial killers like Peter Kürten, the so-called Vampire of Düsseldorf, or the Moors Murderer Ian Brady.[11] Yet cruelty encompasses far more than these rare horrors. Bullying at school and in the workplace, the criticism of celebrities and politicians in the media, and abuse within families can all be viciously cruel, if less spectacularly lethal. These examples are much closer to home; for some they are a painful part of daily life. Few if any of us have never been on the receiving end of some form of social cruelty: the cutting comments or sniggers or sidelong glances which so expertly demolish self-esteem. Few of us, in truth, have avoided being cruel ourselves.

Looking back, we may not much like that nasty former self; but at least we can hope to understand why we were cruel. Social viciousness, verbal abuse, bullying, even perhaps domestic abuse—these we can grasp. They lie within our range of possibilities, in a way in which the Guatemalan killings do not. Surely, therefore, the latter must be peculiarly evil: abominations perpetrated by madmen or monsters. Thus we push the nightmare away, lest it pollute us.

Two claims about cruelty

In this book I will argue in favour of two claims widely accepted by researchers who work on human harm-doing—and often, it seems, disregarded by society as a whole. The first is that cruelty is not, by and large, the domain of madmen or natural-born evildoers. Rather, much cruel behaviour is rational, that is, it is done for reasons which seem good to the perpetrator at the time, and done by people like you and me. Even in the most extreme cases, perpetrators generally know exactly what they're doing. Some, long after they have had the chance to rethink their views, hold fast to the reasons which motivated them to act. Laurence Rees's book on Auschwitz cites a Second World War veteran who 'had helped the Nazis shoot Jews in 1941 [and] still thought he had done the right thing 60 years ago'.[12] Irrational violent harm-doing does occur, for instance in homicides committed by severely psychotic individuals, but on reflection we are reluctant to describe such killers as 'cruel', precisely because their reason is so disordered. Cruelty implies deliberation, free choice, and moral responsibility.

The second claim of this book is that the difference between someone hurling verbal abuse at an immigrant and someone beating an immigrant to death is a difference of degree, not a difference in kind. This is not to say, of course, that the two are the same. Clearly, a mourner fighting back tears at a funeral is very different from a mourner sobbing uncontrollably—yet both are displaying the same type of emotion. We can imagine a continuum, from little or no grief to soul-destroying anguish, on which we could place the two bereaved.

Similarly, we can imagine a continuum of cruelty, from the mildest thoughts and behaviours to the most extreme.[13] At one end lies the initial separation of Them (the Other, the inferior outgroup) from Us (the superior ingroup). Its minor implications include stereotypes, prejudices, off-colour jokes, and mild verbal abuse directed at outgroup members. Moving along the continuum we reach more vigorous verbal abuse, denigratory propaganda, hostile and aggressive stereotyping, the various forms of social ostracism. We find rumours of crimes and atrocities committed by outgroup members, then increasing physical violence against them, their property, and their symbols. Eventually we reach the spectacular rarities which involve destroying people identified as Them, sometimes simply because of who they are, not for anything they may or may not have done.

This continuum of spiralling hostility has been given many names: among them 'demonization', 'social death', 'dehumanization', and 'objectification'.[14] I will use the term 'otherization', which expresses the sense of creating an increasingly impassable social gulf between Us and Them while avoiding the religious overtones of 'demonization', the all-or-nothing connotations of 'social death', and the questions raised by 'dehumanization' and 'objectification' (what is the result of such processes? Are victims seen as animal or mineral, object or agent?). Otherization, like grief, is made up of numerous mental and physical processes inflected by genetic heritage, personal background, and social circumstance. Individuals vary in their susceptibilities, but some degree of susceptibility is part of what makes us human.[15] What this means, in terms of cruelty, is that distancing ourselves from extremes of atrocity, comforting though that response may be, is misleading. The gap between solid citizen and evil perpetrator may be reassuringly wide for most of us most of the time, but it is by no means unbridgeable.

Otherization as a continuum: sliding into the essence trap

> The missionaries of Christianity had said in effect: You have no right to live among us as Jews. The secular rulers who followed had proclaimed: You have no right to live among us. The Nazis at last decreed: You have no right to live.
>
> (Raul Hilberg, *The Destruction of the European Jews*)

Otherization is grounded in the general human bias towards pleasure and away from pain. We prefer and seek out events, actions, and self-descriptions which make us feel good and avoid those which we experience as unpleasant, aversive.[16] Unfortunately, this can lead us to what the writer and student of human folly Charles Mackay called an 'undue opinion of our own importance in the scale of creation', as we overestimate our own wonderfulness and underestimate other people's, ignoring painful truths about ourselves and emphasizing painful truths about them.[17] One way in which we do this, known to psychologists as the fundamental attribution error, is by varying our judgements of whether a particular behaviour is characteristic of a person, or a one-off response to the situation, dependent on whether we like that person or not and whether we approve of the behaviour.[18] For us (and people we care about), outcomes we approve of (like getting a job) may be put down to our own splendid character (impressive qualifications, capacity for hard work, etc.), while negative outcomes (like not getting a job) are down to misfortune or the malice of others (feeling unwell, biased or stupid interviewers, etc.). This rosy tint, like government, protects us against the uncomfortable idea that we might be responsible for the unpleasant consequences of our actions. It gives both groups and individuals a moral discount in their own eyes, protecting them from bruising self-blame.[19]

For people we don't like, however, the picture is reversed. Our bad luck is their deserved come-uppance. The unintended harm we do, if done by them, would be a sign of their malevolent character. By attributing nastiness to the individuals themselves, to intended malice rather than to the situation, we fall into what I will call the 'essence trap'. This involves imagining that everyone has a core character, the essence of who they are, which governs most behaviours but not all, and which is much harder to change than patterns of behaviour.[20] When we do something unpleasant, we put it down to error or misjudgement: a behaviour, easily corrected (or so we assume; in fact behaviours can be very hard to change). When people we dislike do the same, we ascribe it to their unpleasant character, making them seem less changeable (less morally redeemable) and more responsible for their nasty behaviour. In doing so, we emphasize their difference from us, pushing the unpleasantness away to a more comfortable psychological distance and pushing the person away with it.[21] An otherized human being is one whose complexities we no longer notice: a cartoon villain, ripe for persecution.

A pernicious aspect of the human species is our tendency to otherize not just in times of great crisis but in reaction to quite minor challenges to our social status, assets, or sense of honour. Most otherization is relatively minor, but for adults at least some level of otherization is generally the default setting for social interaction: we may have been taught to do unto others as we do unto ourselves, but who among us consistently obeys that principle? More pernicious still, we otherize—sometimes to the point of genocide—when little direct conflict actually exists, purely because we have beliefs about other people which lead us to push them into hated outgroups. We need never have met them and may know little about them, except from sources whose impartiality we should have good reason to doubt; yet we may become prepared to kill them, in appalling ways, because of the beliefs we have about them and the motives which drive us to do harm.

The problem of cruelty shows us otherization at work. To take an extreme view, if cruelty is a behaviour which emerges under certain conditions, only chance prevents us becoming perpetrators of atrocities. Christopher Browning's book *Ordinary Men*, which describes atrocities committed by German policemen in Eastern Europe during the Second World War, reinforces this point (more or less successfully, depending on how happy you are to call yourself ordinary).[22] You too could be a killer, like one of those ordinary people who lived in Nazi Europe (or Bosnia, Rwanda, or all those other ordinary places where massacres happen) who went out one day, or sometimes day after day, to murder their ordinary neighbours.[23] You too could be cruel. Given the circumstances, your cruelty might even attain the gargantuan murderousness of a Tamerlaine, Vlad Drakul, or Saddam Hussein.

Rather than face that possibility, and the moral shallowness and lack of power which it implies, we go to the other extreme. Instead of viewing human cruelty as a behaviour in which even good souls like us may occasionally indulge, we regard it as a built-in flaw of evil people's nature, as inescapable as the need to breathe. In doing so we distance it from ourselves, making it something that only other people do.

A person who *acts cruelly* may have done so for reasons with which we can sympathize, without condoning the action. Such a person could still be considered 'one of us'. If Euripides' Medea or Aeschylus' Orestes, both of whom murder members of their family, were wholly unsympathetic characters then *Medea* and *The Oresteia* would not still be mesmerizing audiences: we see them as human, despite their atrocious behaviour.[24] A person who *is cruel*, however, is unlike us because they are irredeemable; their cruelty is as reflexive as a snake's bite. Asking for reasons in such a case is pointless, you might as well go straight to the DNA. Perpetrators are a breed apart, fit only to be loathed and excoriated. We resent and resist attempts to understand them.

The atrocity stories which circulate in wartime show us how easy and attractive otherization is. These stories feature extreme, senseless, and often implausible cruelty, usually with little or no attempt to explain *why* the perpetrators acted as they did. One example is the tale of an officer who 'had broken into a flat, tied up a man and raped his wife before the man's very eyes. Afterwards the Commissar had literally butchered the wife to death, cut out her heart, fried it in a pan and had then proceeded to eat it.'[25] What were the officer's motives for this extraordinary crime? We are not told. Given how readily human beings usually reach for explanations, the lack of any attempt to use the language of motives and agency is startling, until you see what the story is meant to do.[26] A taste for human flesh is not a common affectation, and yet this anonymous, sketchily presented officer could be anyone; he represents an entire class of people. Is it really plausible that they should all be cannibals? Of course not. Yet that, and hence their irredeemable depravity, is the story's take-home message.

Generalization, misdirection, and obfuscation are useful tools when otherizing foes. An explicable enemy risks being one whose point of view we might come to accept as reasonable. An incomprehensible foe brings no such moral burden. Incidentally, the cannibalism story was told by a Lithuanian to a German in 1941, and the officer was identified as a Jew. Atrocity stories circulate on all sides.[27]

The essence trap is psychologically satisfying. It keeps out the dirt and the danger, preserving our pleasant self-image as justified people who act from reasonable motives. The downside, of course, is that the dirt and the danger are conceptualized as a malevolent, unchangeable agent: a devil whose only purpose is to destroy. This shuts down any attempt at understanding, presenting cruelty as something other than human, an essentially evil force on which our explanations can get no purchase. Viewing cruelty as part of an unchangeable human nature leaves us with only two options: put up with cruel people or remove them.

The usefulness of evil

> And now, what will become of us without barbarians?
> Those people were some sort of a solution.
>
> (C. P. Cavafy, 'Waiting for the Barbarians')

As David Frankfurter notes in his book on myths of evil conspiracy, 'evil', like 'cruel', is used to distance its target from the self or the ingroup, pushing away entities we do not want to resemble from those who are like us.[28] Calling someone or something evil is not like calling it red or slow or sleepy. The term carries crucial moral implications. In effect it makes the following claims:

1. This entity is hostile and threatens our existence and identity.
2. This entity is beyond our powers of influence or negotiation. Nothing we could do could make it change its behaviour.
3. This entity is utterly unlike us. We cannot approach it with kindness or empathy, any more than we would a plague or an earthquake; that would be at best useless, at worst lethally dangerous. Nor can we explain its behaviour as we would our own.
4. This entity is not, like a plague or an earthquake, an inanimate force of nature (though it may be a force of supernature). It is an agent which chooses to act as it does because it desires to hurt, damage, and destroy. It can thus be held morally responsible for its actions.
5. Because the threat it poses is so severe, we are justified in neutralizing that threat by any means possible.

Calling someone 'evil', Frankfurter argues, creates 'an implicit boundary beyond which acts are no longer worthy of context or empathy'.[29] It deters attempts at explanation by rendering them morally suspect. Even in thought, crossing the boundary from Us to Them can be regarded as a threat.

We must be careful, however, to distinguish empathy from agency. The person we call evil is someone we wish to have nothing in common with; even the process of empathizing with them might make our thoughts and feelings dangerously similar to theirs. Yet although we cease to empathize, we do not thereby remove all the assumptions of folk psychology. In particular, evil people are still agents. They remain responsible for their actions, and hence within our moral universe (they are just relocated to its undesirable neighbourhoods).[30] In fact, their agency and responsibility is greater than ours. Why? Because we can always find reasons why we did what we did, excuses which offer (we hope) a moral discount on our bad behaviour. But evil is incomprehensible; there are and can be no reasons, no moral discounts. At the same time, evil is an active malevolent force, which can be blamed for its joy in doing damage without qualms or mitigating factors, thereby justifying revenge. It is a truly free agent—in a way that we humans, tangled as we are in the chains of causality, can never be. Evil people can be hated, blamed, or annihilated without guilt, since they are no longer truly human.

This gives us two ways to otherize our enemies. The first uses language like 'malevolent', 'treacherous', and 'cunning', speaking of witches, spies, and secret agents, devils and demons, and enemies. All of these fully intend the harm they do. The second refers to 'deadly', 'venomous', 'contaminating' forces which inflict sickness, weakness, and death: dangerous animals, tumours, germs and their carriers—all of which destroy unthinkingly, because it is their nature.[31] Inconsistent as it may seem to imagine your enemy both as agent and non-agent, actively malicious and impersonally destructive, the language of atrocity employs both forms of otherization.

And it works, despite the contradiction. Not many perpetrators seem to notice the problem, and those who do can easily dismiss it as 'just metaphorical'. Perpetrators are usually motivated to accept otherizing statements without much pedantic conceptual analysis (unless they happen to be located in a novel by Sade, in which case intellectual challenges are welcome). On a practical level, once the belief that the enemy is unreasonable and unchangeable has been accepted, it doesn't really matter whether the evil one is attacking out of pure malice or acting instinctively according to its nature. Either way, force will still be needed to stop it.

For those who use hate propaganda for political purposes, having both forms of otherization actually makes the propaganda more effective. Implying that an enemy fully intends to do you wrong places that enemy firmly within the moral universe, allowing you to justify your behaviour as self-defence or righteous punishment (i.e., not cruelty). Meanwhile, implying that an enemy is by nature destructive, like cancer, encourages appropriately extreme responses. Cancerous cells may be part of the human being they afflict, but that affiliation does not deter wholehearted attempts to eradicate them. Moral qualms can thus be assuaged by talk of evil enemies, while action is encouraged by emphasizing the more inanimate view of them.

The journey ahead

Thinking of cruelty as largely rational behaviour carried out by people as human as we are can be deeply uncomfortable. Yet it does allow us to ask about the reasons why people act cruelly. Understanding those reasons, the first step in changing behaviour, is undoubtedly more difficult in the short-term than tolerating—or removing—those who are cruel. Understanding, however, may offer a more effective long-term solution to the problems raised by human viciousness. Viewing extreme cruelty as related to and emerging from less serious forms of harm-doing enables us to examine the factors which push us along the otherization continuum. Thus we can begin to outline a scientific approach to cruelty.

The study of cruelty is a fraught and contested enterprise, and scientific approaches to cruelty risk many misunderstandings. First of all, I should make clear that I am not suggesting that science has *the* answers and that scientists can therefore ignore research from other disciplines. The concept of cruelty is not scientifically coherent in the way that the concept of a proton is coherent. Cruelty is dependent on social—moral and cultural—variables as well as on physical aspects of how human beings are built. What scientific approaches can contribute is dissection of the concept of cruelty into its components, some of which can be fitted into pre-existing scientific frameworks, providing general (e.g. evolutionary) explanations for harm-doing. Arguments for why perpetrators can come to enjoy hurting and killing people, however, do not explain why some murderers douse their victims in petrol and burn them, some bury them to waist-height and then set dogs on them, and some shoot them. Understanding particular atrocities requires grasping the local context at the time. In other words, science working in concert with philosophical, anthropological, and other approaches may achieve much more than any one discipline in isolation. This book is thus, among other things, a plea for more interdisciplinary research.[32]

A second, related pitfall concerns the specific science whose findings will be central to this book: neuroscience, the study of how brains work. Can it offer us anything deeper than descriptive 'geography'— the fMRI stereotype of colourful pictures based on difficult statistics?[33] I believe it can; neuroscience is more than fMRI (and fMRI is more than just images). Brain research can give us causal mechanisms, answering questions about how cruel behaviour happens. When combined with the adaptationist perspective derived from Charles Darwin, it can even address the question of why we are cruel. Neuroscience is a young endeavour, and there is much we still do not know.[34] Youth, however, does not imply impotence.

There is also a wider criticism, often made by social scientists, of the scientific study of violence. Just as explaining all cakes with a single recipe—butter, sugar, eggs, and flour; combine as follows—says nothing about why cakes are ginger, chocolate, or vanilla, so explaining all violence in terms of brains or genes is impossible, because violence is socially variable. Two very similar groups may show different patterns, and violence can erupt, or ebb, over very short timescales.[35] So what makes the difference? It cannot be biology or genetics, since these neither change that quickly nor vary that much between people. Right?

Wrong. Take the case of genetics. Genes may pass on inherited characteristics, but once they have built a new organism they are not then unemployed until it has sex. Genes in the brain and body switch on and off all the time, contributing not only to differences between people but within a single person from day to day. Brains too are highly variable, both structurally—some adult brains lack major anatomical landmarks, for example—and functionally. Much of what they do, moreover, goes unnoticed by the person or by other people. Aggression, however sudden, never erupts without prior preparation in the brains of those who perpetrate it—yet until the eruption their actions may appear quite unremarkable. Thus extensive brain changes may occur undetected, allowing apparently minor causes to trigger ferocious violence. If explaining violence and cruelty purely in biological terms is like leaving out what makes a cake distinctive, explaining them purely in social terms is like offering a recipe which makes no mention of flour. Both approaches are required.[36]

A note on terminology is also needed here, because this book's title is neither *Violence* nor *Aggression*, but *Cruelty*. The three concepts, all highly contested, cover the same terrain of hostility and harm, but there are differences. 'Violent' describes behaviour, calling to mind impressions of rapid action, physical force, and damage. 'Aggression' speaks more to the hostile intention; one can be highly aggressive without actually hurting people. Cruelty often involves extreme violence, yet it need not even touch its victims. One could expand the concept of violence to include verbal cruelty and cruelty through neglect or withholding of benefits, but that is quite a stretch of everyday usage.

As for aggression, researchers commonly divide it into two kinds: reactive and instrumental. Reactive aggression is often perceived as rapid and uncontrolled, a 'lashing out' response to frustration or provocation which can be seen in infants. Instrumental aggression is more deliberate, and may lack an obvious provoking cause. The two differ psychologically and neurologically. People with high reactive aggression often grow out of it and are not necessarily unpleasant, cruel, or criminal. Individuals who plan their aggression, on the other hand, typically show callous and unemotional traits from childhood. Nowadays these infant psychopaths are often diagnosed with conduct disorder by adolescence and full-blown psychopathy by adulthood.[37] They are more likely to be cruel to animals as children and to commit violent crimes as adults—or sometimes before they reach adulthood.[38]

Cruelty, with its deliberate, hostile exploitation of others, is closer to instrumental than to reactive aggression. What neither concept of aggression captures, however, is the context of cruelty: the moral and social biases which lead to people or their actions being called cruel. Given the wealth of material already available on violence and aggression, therefore, the focus of this book will lie elsewhere, at the interface between scientific and social perspectives on cruelty.[39]

A final caveat concerns the scope of the ideas set out in this book. Cruelty means many different things. Can any one approach really encompass the range of people's cruel behaviour, from boys dismembering invertebrates to men dismembering boys, or even babies? In principle, a complete scientific 'theory of cruelty' should do exactly that. What *Cruelty* offers, however, is necessarily incomplete: preliminary groundwork on a theory of cruel behaviour. I have focused on extreme cruelty. The ideas I propose may also be applicable to less dramatic forms of viciousness, but constraints of space prevent their consideration here.

Cruelty is structured around nine questions. The first three chapters ask 'what', 'who', and 'why': what is cruelty, who decides what counts as cruel, and why do humans have this unpleasant capability? We will look at the social context of cruelty and the roles of the three key participants—perpetrator, victim, and judge/observer. Cruelty is clearly a moral concept, so we will consider the topic of morality in some depth before moving on to the evolutionary forces thought to provide the 'deep reasons' why we are cruel.

Chapters 4 to 6 dig deeper, enquiring into the biological mechanisms involved in a person's decision to act cruelly. That means contemplating brains in necessary detail, since brains are the common medium through which every causal factor involved in cruelty, from genes to drugs to social expectations, influences patterns of cruel behaviour. Chapter 4 looks at how we decide to act, Chapter 5 at how we come to feel the emotions which motivate our actions, and Chapter 6 at the beliefs which guide them.

Finally, Chapters 7 to 9 investigate kinds of cruelty. Chapter 7's topic is the callousness which disregards its victims; Chapter 8 discusses the sadism which savours their agony. In both cases we will look at motivations and mechanisms. Chapter 9 summarizes the themes of the book and addresses the issue of whether we can learn to stop being cruel.

Summary and conclusions

At times, as in Eastern Europe under the Nazis, cruelty seems to become almost normal, the default, indiscriminate response of twisted people in a deranged society. But the impression is mistaken. Cruelty does emerge in certain situations, but never indiscriminately, except perhaps in the cases of severe mental disturbance where someone 'goes berserk' or 'runs amok'. Even when a mob or army sweeps through a town, intent on massacre, the killers discriminate: some townsfolk may be spared, and for all the talk of blood-lust and frenzy perpetrators in crowd violence rarely kill each other. Calling a mob demonic may be tempting but is itself a form of otherization, not only simplistic but dangerously misleading. The essence trap is no more applicable to a mob than it is to a single perpetrator, however vicious.

If cruelty were purely a product of human character, the result of a few 'bad apples' with bad genes, then in principle it could be wiped out, either by murder or by sterilization. Many utopian visions have inspired just such campaigns of killing to remove 'flawed elements'. If, however, cruelty is affected by circumstance as well as by character, then people will continue to be cruel in certain situations. Killing them will not stop cruelty (as utopians learn, once you start, you have to keep killing, as your cruelty breeds cruelty in response). Only changing the circumstances will do that.

The dichotomy—nature versus nurture, character versus circumstance—is of course oversimplified. Both, as we shall see, play a role in vicious behaviour. First, however, we must begin with the basics, and ask what it is we mean by cruelty.

Chapter 1

What is cruelty?

In the summer of 1941 a German army photographer was sent to the Lithuanian city of Kovno (now Kaunas):

I was confronted by the following scene: in the left corner of the yard there was a group of men aged between thirty and fifty. There must have been forty to fifty of them. They were herded together and kept under guard by some civilians. The civilians were armed with rifles and wore armbands, as can be seen in the pictures I took. A young man—he must have been a Lithuanian . . . with rolled-up sleeves was armed with an iron crowbar. He dragged out one man at a time from the group and struck him with the crowbar with one or more blows on the back of his head. Within three-quarters of an hour he had beaten to death the entire group of forty-five to fifty people in this way. I took a series of photographs of the victims. . . .

After the entire group [had] been beaten to death, the young man put the crowbar to one side, fetched an accordion and went and stood on the mountain of corpses and played the Lithuanian national anthem. . . . The behaviour of the civilians present (women and children) was unbelievable. After each man had been killed they began to clap and when the national anthem started up they joined in singing and clapping. In the front row there were women with small children in their arms who stayed there right until the end.[1]

What is cruelty?

Viciousness has an ancient genealogy. 'Cruelte', which is recorded in the *Oxford English Dictionary* before AD 1225, is thought to derive from the Latin noun *crudelitas*.[2] That seems appropriate: it is from Rome, particularly Imperial Rome, that some of our most impressive examples of cruelty originate. Think of the hideous suffering of crucifixion, of people used as human torches in the emperor Nero's gardens, of the gladiatorial games.[3] But cruelty far pre-dates Rome. Here, from Homer's epic war poem *The Iliad*, is King Agamemnon in the heat of battle, raging at his brother Menelaus for thinking of sparing a Trojan's life for ransom—words thought to have been composed at least twenty-seven centuries ago:

> So soft, dear brother, why?
> Why such concern for enemies? I suppose you got
> such tender loving care at home from the Trojans.
> Ah would to god not one of them could escape
> his sudden plunging death beneath our hands!
> No baby boy still in his mother's belly,
> not even he escape—all Ilium blotted out,
> no tears for their lives, no markers for their graves![4]

What do we mean when we call a person or an action 'cruel'?[5] Wartime atrocities like those in Kovno provide paradigm cases of cruelty; but what makes us recoil from the thought of that massacre and readily call its perpetrators cruel? One factor is surely the victims' suffering. This may be not so much physical—there are worse ways of dying than being hit over the head—as psychological. We quail at the agonies they must have felt, having to wait and watch, perhaps as people they loved were killed in front of them. Knowing death was inescapable, seeing their neighbours cheering on their murder; that we judge to be psychological torture. Empathy can bring us no more than a faint shadow of their anguish; nevertheless, it suffices to distress us, giving emotional weight to our moral judgements.

We can also consider these events not from a victim's perspective, but from that of an onlooker or even the perpetrator. This can evoke intense horror and disgust: we imagine blood spraying, cracked bones, the sights and sounds and stenches of mass murder. This can serve to distance us from the perpetrator, appalled as we are by the idea of creating such carnage. But disgust is double-edged: it can also push us away from sympathy for mutilated victims.

Another disturbing factor is the obvious rejoicing of the perpetrator and the complicit onlookers. They don't strike us as coerced underdogs, forced to butcher by their savage German masters, but as delighted volunteers enjoying the destruction. If there is compulsion here, we don't see it. Instead, their participation seems freely chosen, their murderous intention obvious.[6]

Two further influences also affect our judgements. One is justification, or more precisely, the lack of it. To our eyes, the murders seem unjustified, gratuitous.[7] Even if the victims had been communist spies or supporters of Russian oppression, could they not have been put on trial like other criminals? Of course this was wartime; but even wars have codes of conduct, formalized in the Hague Treaties and Geneva Conventions decades before the Second World War.[8] Prisoners of war may be detained for the duration of combat; spies and terrorists locked up until they can be brought to justice. Mass killing seems unnecessarily brutal, given the other options which should have been available. In the case of the Kovno massacre, we are motivated not to look too hard for justifications, or to accept those we come across, by our pre-existing bias against the perpetrators: Nazis are evil, everyone knows that. Had we been considering an outrage committed by our own armed forces the search for justification would have been much more extensive.

Complementary to the issue of perpetrator justification is the issue of victim innocence.[9] The motive publicly claimed for the Kovno massacre was revenge: an oppressed people striking back against their Jewish-Bolshevik oppressors. The German officers who were present but refused to intervene used the claim of retaliation as an excuse, saying that a purely internal conflict was not their responsibility. We, however, see the victims as innocent Jews murdered because of their ethnicity, irrespective of their actions or allegiances. Perhaps some of the dead had previously assisted the Russians. Yet even if we were to accept that guilt we might still judge their 'punishment' disproportionately cruel.

To summarise, the major factors which prompt us to define behaviour as cruel centre on the two axes of the participants involved: the *perpetrator*'s motives and behaviour (unjustified/gratuitous, voluntary, intentional) and the *victim*'s status (innocent, undeserving) and experience (suffering). A third axis, of course, is the *observer* who makes the moral judgement. We can bring these factors together to say that in everyday usage 'cruelty' involves *unjustified voluntary behaviour* which *intentionally* causes *suffering* to an *undeserving* victim or victims. In addition, cruel behaviour often, though not always, affects observers, evoking disgust directed at perpetrators and empathy (and sometimes also disgust) for victims. This is a working definition, if you like, of cruelty: an attempt to capture how the concept is used by the general public. It will need modifying, but it gives us a starting-point.

If these conditions do not apply, we tend to label the behaviour in question as something other than cruelty. If the victim deserves what happens to him or her, we view the perpetrator as meting out due punishment. If no suffering occurs (for example, if pain is caused but the victim appears to enjoy it, as in masochism), then the perpetrator may be deviant, may even be said to have cruel intentions, but his actions do not qualify as cruel, nor is his target a victim in the usual sense. Suffering caused accidentally or unintentionally may be negligent but is not cruel, while coerced harm-doing transfers the charge of cruelty from the person who acts to the person who forces them to act. Finally, behaviour may encompass both action and deliberate inaction. A person who causes another to suffer may do so by actively hurting them, or by withholding the means of relieving their suffering.

Cruelty is clearly a moral concept, closely related to other moral concepts such as punishment, justification, and responsibility. We shall look more closely at morality in the next chapter. First, however, we need to consider the components of our definition in more detail, beginning with the concept of suffering.

Cruelty is behaviour which causes suffering

Cruelty, first and foremost, is doing harm. Dictionary definitions make it clear that a cruel act is one which causes pain or suffering, in which the perpetrator is at best indifferent to, at worst gratified by, the suffering he or she causes. The *Oxford English Dictionary*, for example, defines cruelty as a 'disposition to inflict suffering; delight in or indifference to the pain or misery of others; mercilessness, hard-heartedness: *esp.* as exhibited in action. Also, with *pl.*, an instance of this, a cruel deed.'[10] Note how the definition fluctuates between action and disposition. Cruelty is about qualities of character as well as behaviour; persons as well as their actions are labelled 'cruel'. Behaviour, however, comes first. The label 'cruel' is extended from action to actor primarily on the basis of behaviour, and only later, given appropriate stereotypes, to other aspects of the person (for example, their savage appearance, fierce expression, or vicious reputation). We can extend the concept of cruelty within the skull—to thoughts, desires, and intentions—but even these have to be articulated, whether internally by us or externally, expressed in action or speech. Only then, having been given life by some action, can they be judged and labelled cruel.

To be so judged the behaviour must cause suffering—and that requires a sentient victim capable of experiencing it. Robots as traditionally designed are unable to suffer and cannot be victims of cruelty, any more than computers can. Animals can, to varying degrees; insects and arthropods we're not so sure about; bacteria, who cares? As far as we know, the world's number-one expert when it comes to suffering is *H. sapiens sapiens*. That goes for both giving and receiving.

Suffering, as the example of masochism shows, is not the same as feeling pain. Suffering involves both harm (including physical pain and psychological distress) and aversion, which guides us to avoid its causes in future. A victim must experience harm *as unpleasant*. Individuals given opiates for pain relief distinguish pain and suffering when they say that the pain is still there but it doesn't matter so much.[11] The drugs reduce their aversion to the pain, and consequently the degree of suffering they experience. Pain 'in a good cause', for instance the ache of surprised muscles after the first session of a new exercise regimen, may be tolerated or even welcomed as a sign of achievement.

Some people cannot feel pain at all, thanks to a genetic mutation, but they can still suffer.[12] If they were threatened with death, for instance, suffering could occur for many reasons: the fear of death, the thought of the pain that will be caused to loved ones, grief at the loss of future potential, and so on. Human beings, and perhaps some other species too, can suffer torments without experiencing physical pain, as anyone knows who has seen—or been—a person plunging into depression, or breaking up with a partner, or suddenly bereaved. Physical pain is not essential to cruelty. Social viciousness, which never so much as bruises the victim's skin, can cause overwhelming emotional distress. A cruel action is thus one which causes suffering: unwelcome, aversive pain, distress, or harm.

Cruelty is voluntary

Judging an action as cruel is frequently a precursor to moral condemnation and punishment of the perpetrator, whether punishment is inflicted in order to change their behaviour or because of the desire for vengeance. Just punishment, however, is applied only to morally responsible perpetrators: agents who act freely, cause the consequences for which they are being judged, and choose to harm others.[13] We regard a mentally ill person unable to understand what he or she has done as requiring treatment rather than punishment, while a person forced into cruelty by, say, the threat of imminent death is generally held to be less responsible for his or her actions than one who freely chooses to be cruel. The precise categorizations of 'responsible' and 'not responsible' vary across and within cultures, over time and between different groups (a psychiatrist and a jury member in the same courtroom may make very different judgements about someone's responsibility for his actions). The categories themselves, however, are cultural universals, like 'good' and 'evil' or 'pure' and 'impure'. All over the world, people seem to assume that they and other people have some ability to choose what they are going to do next.

Agency is that power of voluntary choice: the level of control over and ownership of one's actions considered typical of healthy adult humans. Agency gives us the ability, providing we are not impaired by drugs or illness, to act on and change the world in order to satisfy our desires and needs. It gives us causal independence, in that by acting we make things happen which otherwise would not have happened. At least some of our actions occur because we want or intend them to. We, not chance or external causes, make them happen.

Human beings naturally use the language of agency. We think in terms of beliefs and desires, intentions and reasons to act. We explain what people, and often animals, do in agent-based terms: 'he wanted to help her'; 'they thought the rabbit was lonely'. Indeed, we are so used to setting events in an agent-based framework that we see purposeful behaviour where it does not in fact exist. For most of human history people have ascribed natural disasters to irate deities. We find faces in clouds, see car headlights as eyes, or attribute goals and intentions to small moving shapes on a screen.[14] Evidence from social neuroscience suggests that we are deeply biased towards interpretations in terms of agency. From the start, our brains are specialized to detect and analyse facial expressions, gestures, and the patterns of movement made by living things.[15]

Thinking in terms of beliefs and desires, of agents who cause events and act for reasons, seems not to be restricted to our own species. Causal responsibility, of course, goes with being alive: wolves cause caribou to die because they want to eat them; salmon force their bodies upriver to breed; even trees can be said to be causally responsible for the leaves and seeds—and cracks in the paving—they produce. Trees, however, are not thought to have much understanding of causal relationships. More complex species, by contrast, go far beyond the mere capacity to cause events, acquiring sophisticated causal knowledge of likely outcomes— and, crucially, the ability to incorporate knowledge about those out-comes as reasons for acting, or not.[16] Humans are particularly good at this ability to take account of causes and effects. We predict events and change our behaviour accordingly. We also predict our behaviour, and that of other people, using agent-based thinking. We guess at their beliefs and we infer their motives.

Cruel behaviour is judged by reference to motives and intentions

Formulating cruelty as unjustified behaviour which intentionally causes undeserved suffering emphasizes the importance of how motives for acting are interpreted—by observers as well as perpetrators and victims. Motives are clearly a crucial issue in judgements of cruelty. Accidental harm, or harm due to natural forces, may be devastating, but it lacks the necessary motivation; so much so that we often supply the deficit metaphorically, describing the victim's misfortune as 'cruel fate' or 'cruelly unlucky'. Faced with a judgement about another's harm-doing, our first question is often 'why did they do it?' We look for explanations, using the agency framework: the language of beliefs and desires, of purposeful actions and motives.

Assessing the motives of others can at times be easy (we know at once that the angry man approaching us wants us off his lawn), but in cases of cruelty it can be difficult or impossible. Perpetrators may refuse or be unable to explain themselves; they may claim to have no idea why they did what they did.[17] As we shall see, there is more to this than the uncooperativeness of the naturally vicious or the heavy use of drugs and alcohol. For one thing, movements consciously begun can trigger more instinctive behaviour which is often so rapidly enacted that it feels automatic. Young men who chase a victim intending to scare him, for example, may find that their brains reinterpret the chase as a hunt—and treat the victim as prey accordingly.[18]

Motives can also change with confusing speed. A man who stabs an opponent in a fight may have begun the fight in response to an insult, with the goal of defending his honour. When his opponent did not back down the goal became 'to teach him a lesson'; when his opponent grabbed a weapon the goal became physical survival. Which motive led to the stabbing? Presumably all three. Single-motive actions are easier to interpret, but many actions have multiple contributing motives, just as one motive can influence more than one action.

Imagine you are serving on a jury in a murder trial. A woman admits that she killed the victim, but claims that she did so because he was about to rape her. Was he? What's the evidence? Was she bruised; were there neighbours who heard screams; had she complained of threats from him before? This is your task: to try to sieve out the facts of the case—including the psychological facts of the defendant's likely motives. If her attacker was a stranger, you can use your accumulated experience of how people behave to infer that her motive was probably self-defence. Of course, she might have been brainwashed by a cult into such a ferocious desire to kill people that she chose a man at random and made up the story of his attack on her. She might also be a predatory alien; but neither of these options is common in the world which shaped your perceptions. Men attack women, however, all the time, so until you are given countervailing evidence self-defence is likely to be your baseline guess about the woman's motive.

In practice, motives are frequently mixed, as well as hard to define and sometimes opaque to their possessors. If your defendant, for instance, knew her killer your work as a juror will be that much more complicated. What is the truth of the history between them; how did she feel about him; did she have reason to want him damaged or dead? Perhaps they had been a couple, and she left him because of his violence, or he left her because of her continual lying. There may be evidence for both claims and not enough evidence for either to win out. Rape cases in particular illustrate the limitations of our ability to get to the truth about motives, which is one reason why conviction rates remain so low despite attempts to raise them. Rape is undoubtedly cruel, but sex in itself need not be. If we have no evidence that the victim suffered physically, we must attempt to establish both parties' motives by asking ourselves, in effect, how we and people we know would have behaved. That kind of knowing can range from direct experience to familiarity with cultural stereotypes: rapes portrayed in fiction, for instance.

Our initial definition conceived of cruelty as involving intentional behaviour. People who act cruelly must, we feel, have intended as well as wanted to do so. All of us have desires we never intend to fulfil. Philosophically, issues of motive and intention take us into very deep waters in the search for a definition of (in this case) cruelty which can precisely distinguish what is cruel from what is not. The high road of abstraction, however, is not the one this book will take. Focusing on how the concept of cruelty is used in everyday life, we can set the demand for precision to one side.[19] Instead, let us think of cruelty as more-or-less rather than yes/no, a cluster of several component ideas which we can group around the two axes of perpetrator and victim. These components each contribute to our judgement of whether the act is cruel. All of them may be more or less clearly present in any given case of potential cruelty, and they can be combined methodically or intuitively, consciously or instinctively, in any order or all at once. The clearer and more numerous they are, the more likely we are to describe the behaviour in question as cruel.

Victim-related components, as we have seen, include the issue of whether the suffering is deserved. Victims may be innocent in general (e.g. babies) or specifically innocent of the wrongdoing cited as justification by perpetrators (for example, when the justification is of that common and ludicrous variety, 'your ancestors hurt mine so now I'm hurting you'). Victims may even be guilty and deserve punishment. Their guilt, however, is a separate question from whether their punishment is appropriate or, in the words of the US Constitution, cruel and unusual. Consider the following three perpetrators of torture. One burns his baby stepson with a cigarette. The second, adopting Old Testament principles of retribution, burns a man who burned his baby stepson, with a cigarette. The third modifies the 'eye for an eye' approach by using a flamethrower instead of a cigarette (this is inflationary cruelty of the kind practised by the Khmer Rouge, which the anthropologist Alexander Laban Hinton calls 'a head for an eye').[20] The first situation is a paradigmatic case of cruelty, most people would surely agree. There is much less agreement in the other cases, in which the victims' innocence is diminished, although I would expect the third perpetrator to be judged more cruel than the second, since flamethrowers hurt more than cigarettes.

Perpetrator-related components, which likewise combine to ratchet up (or down), the extent to which behaviour is seen as cruel, include a lack of justification, voluntary action, and a third component which relates to the perpetrator's intentions and desires. Motivation both matters and varies considerably. For example, a murderer who says 'I only meant to frighten him' is likely to be judged less cruel than one who claims 'I wanted to see him die screaming for his mother'. For the purposes of the argument, let us assume that we can take these killers' statements as genuine reflections of their motives, that neither is mentally or physically ill, that their victims were not strong enough to fight them, that no one was forcing them to kill their victims, and that at the time of the killing they were alone with their victims and not under any particular threat. Given these assumptions, did the killers intend to act as they did? They were motivated to hurt, but did they intend to kill?

Like a typical academic, I must answer: that depends; because 'intend', like 'voluntary', is a slippery word. For one thing it depends when you apply it. The first killer may not have intended murder when he began the process of frightening his victim. Nonetheless, at some point he acted (voluntarily) in such a way as to kill. Perhaps it was accidental, in which case we can ask whether he could reasonably have been expected to foresee the possibility of such an accident, and take steps to avoid it. Judging cruelty, in other words, involves assessing not only perpetrators' motives but their awareness of consequences. Cruel perpetrators foresee, or should have foreseen, that their behaviour will hurt their victims; they do it anyway. We should therefore adjust our working definition: cruelty is unjustified voluntary behaviour which causes foreseeable suffering to an undeserving victim or victims.

One way to clarify this complexity is to use the concept of action-goals: objects, events, or states desired by an agent. Agents act in ways they believe will help to convert their goals from dreams to realities. The first killer's action had as its main aim the goal of frightening the victim, not achieving his death. The second killer appears to have had at least two goals: to murder his victim and to make him suffer as he died. Actions can have more than one goal, and goals may require many actions to achieve them. Goals can also vary in their importance to the agent, both with respect to other goals and over time, as in our earlier case of the man getting into a fight. This is a fluid and compli-cated vision, of more-or-less inchoate goals—some long-lasting, some only momentary—competing for the agent's attention and for control of his or her behaviour.[21] (In Chapter 4 we shall see how this fluidity is grounded in the structure and function of the brain.)

A goal's importance is influenced by how much the agent desires it, and that in turn, as we shall see later, is affected by the agent's experience of how rewarding it is to achieve that goal, or similar goals. A child play-ing with toy bricks (do children still play with bricks in these sophisticated times?) may begin with the goal of building a tall tower; yet he may soon discover that knocking towers down is much more fun than building them. The destructive goal, which had not previously occurred to him, now dominates his tower-building behaviour. Similarly, our two killers could both plausibly have started out with the same goal of inflicting pain and terror on their victims. In one, however, the thought of killing became increasingly desirable, until it became a goal important enough for him to act in pursuit of it. The other killer may or may not have considered his victim's murder desirable; if he did, he did not rank it nearly as highly as his colleague. Likewise for the victim's suffering. One perpetrator inflicts it for instrumental reasons, to frighten his victim. Terror is the aim of the exercise; suffering the method employed. For the second perpetrator suffering, it appears, is both a means and an end in itself. Unlike the first, he acts because he enjoys it.

Two faces of cruelty

The distinction between wanting to achieve a goal because it is desired and wanting to achieve it as a step towards another, desired goal is mirrored in the *Oxford English Dictionary* definition of cruelty cited earlier, which describes cruelty as 'delight in or indifference to the pain or misery of others'. This takes us to the heart of a confusion over what counts as cruelty, because indifference and delight seem very different. The former is ostentatiously neutral, a product of the rational, weighing-up brain; the latter is vividly emotional. Indifference leads us to costs and benefits, to assessing perpetrator incentives and levels of victim suffering. Delight tips us into the moral sphere, because enjoying cruelty is just plain *wrong*.

A young man mugs an old lady to get money for his next drug fix. He is indifferent to her suffering, or at least cares more about his chemical needs than her pain, but is he driven by a delight in inflicting misery? Probably not. Offered a way to get money, or drugs, which didn't cause suffering, he might well take it, suggesting that his primary goal is cash, not torture. The harm he does is a means to an end which to him is hugely important. In assessing how we punish his bad behaviour, we can use reason to weigh up his incentives, his motives for action, and any mitigating circumstances, and set against them the damage done to the old lady. In calling his behaviour wrong, we argue that he should have assigned a stronger moral prohibition to beating people up. In other words, we may think of him as assessing goals and desires, costs and benefits, using the same psychological mechanisms that we use. We the judges do not consider that the ends (financial gain and the consequent satisfaction of a drug-craving) justify the means (terrorizing the elderly). Nevertheless, we accept that although the mugger voluntarily inflicts suffering on a helpless old lady, that is not the main reason why he mugs her. Rather, the attack is a means of achieving his immediate aim (getting money) in order to satisfy his overarching goal of obtaining drugs.

To judge in this manner is to see cruelty as callousness and respond with reason: a search for motives, incentives, and deterrents. We may feel outraged at the mugger, but in general that outrage does not override utilitarian concerns. Our instinct is to punish, but also to correct; we feel that there is hope for the young criminal, if only he can be made to repent. The causes of his actions may lie in his psyche, but they are changeable causes. Ignorance, laziness, impulsiveness, and an immature sense of empathy, as well as his chemical craving, have caused him to rank his values in an antisocial fashion. Once he can be persuaded to revise the list, to view hurting elderly women as unjustifiable, he can become a model citizen. We are not always so calm, of course; cries to 'lock 'em up and throw away the key' bear witness to that. But we have defences against such violent passions in our legal systems, however imperfect. They are designed to foster rational thinking, even in cases of callous cruelty.

And then there are the extreme cases, like the man reported, in a study of cruelty to animals, to have buried a cat to neck level and then used a lawnmower to decapitate it.[22] He took some trouble in pursuit of this revolting goal, and it is hard to explain that effort in terms of any ulterior aim; there are easier ways to fertilize one's lawn. Lacking alternatives, we conclude his motive was cruelty as an end in itself: delight in suffering, the urge to hurt and destroy for hurting's sake. That is, we see his cruelty as sadistic and morally repugnant, and our response is intensely emotional: horror, pity, fury, and most of all disgust. It is hard to be coolly rational, to weigh up the costs and benefits when deciding how best to punish the perpetrator. It is easier to push the very idea of him and his crime away from us, as if it were contaminating even to think of such things. This is the intensity of moral judgement.

The term 'cruelty' implies a moral judgement

Cruelty is unjustified behaviour, inflicted on an undeserving victim—or else it is appropriate punishment. Justification, however, moves us squarely into the realm of morals, of right and wrong behaviour. A victim's injuries can be quantified dispassionately, to some extent, and a perpetrator's aggression confined within more or less scientific metrics. Cruelty, which typically involves both observable injury and obvious aggression, nevertheless has this additional moral component: the reasons for acting are somehow not good enough.

Returning to our example of a murder trial, let us assume that there is considerable evidence for the victim's violent intent and the defendant's consequent motive to defend herself. Your next task is to assess whether that motive was a justified reason for her to act as she did. Here, as jurors, you have an extensive body of law and custom to guide you. Your options are restricted, especially if the defendant has admitted the act of killing. In principle all you have to do is weigh up the reasons, the background details, the mitigating factors, the pros and cons, and from that calculation comes a verdict. In practice, research suggests that juries, like the rest of us, are biased by factors like the defendant's status, appearance, and demeanour. Handsome and personable criminals tend to get off more lightly than ugly unfriendly ones, and stereotypes appear to affect the outcome.[23]

Juries also have another source of bias to contend with, one against which judges habitually warn them. That is their moral revulsion, their horror and disgust at the crime itself. Assessing issues of motive and justification, in other words, is complicated by the fact that we appear to have two ways of judging other people's actions.[24] One is slow and thoughtful: it considers reasons, weighs costs and benefits, and asks for evidence. The other is swift and instinctive: a moral, emotional, immediate 'yes' or 'no'. These two systems appear to vie for our attention, and depending on the circumstances we may favour either or both. Judges may plead for rational decisions, but moral outrage can be hard to set aside.

COOL MORALIZING

When judges urge jurors to suppress their moral responses in favour of more dispassionate reasoning, they are promoting the form of judgement we think of as traditional rationality: analysing reasons and causes, costs and benefits, to arrive at an appropriate decision. This is the framework we use when we are observers, detached from and not directly affected by the events we are interpreting. In its pure form this is the ideal of reason, leisured and abstract, judging without commitment. It may analyse human behaviour without any reference to morality (for example, in the human sciences), or it may incorporate moral claims, as in the courtroom.

Applied to the moral domain, such cool and thoughtful decision-making is often associated with 'utilitarian' thinking, but it need not be as formal—or as unselfish—as the 'greatest good for the greatest number' philosophy associated with Jeremy Bentham.[25] In principle, utilitarianism achieves a lawmaker's (or a god's) neutrality, asking what most benefits most people without assigning any special privilege to the self. In practice, even the chilliest powers of reason are easily biased. Computers may practice cold logic, but human brains, however practised in abstract thought, keep an eye out for self-interest. Motives and intentions, reasons and justifications—we are experts at justification—are evaluated in terms of costs and benefits to the people involved, the wider social group, and the assessor. Liking, familiarity, complexity: these and other biases affect our judgements whether we realize it or not; and because we do not experience them in the way we experience emotions (that is, they are 'cool' and cognitive rather than 'hot' and affective), we tend not to notice their presence. Preferences for the known, the simple, and those people similar to themselves may subtly sway a judge or a philosopher, not just individuals untrained in reasoning.

In describing thought with the analogy of temperature, incidentally, I am adopting a very common metaphor of reason and emotion.[26] Like many favourite metaphors, it is grounded in the body: in this case, the physiological reactions which accompany emotional reactions and prepare the body for fast and strenuous movement. Not every angry person is red-faced and sweating, but that is the caricature. The temperature metaphor also applies to different emotions. I would guess, for instance, that you think of disgust as cooler than rage, of love as warmer than liking, and so on.

On this view, people's cool moral judgements are not simply derived, *pace* Socrates and many philosophers since his time, by the careful application of logical rules. They also weigh up incentives and deterrents, potential costs and possible benefits, acting so as to maximize what economists term 'expected utility'. Of course, individuals may have differing beliefs, knowledge, preferences, or biases which lead them to vary in how they value their options (a nightclub might be your weekend pleasure and my idea of hell), but the process of weighing up is much the same: the rational choice beloved of economists. (Or, more realistically, semi-rational choice, given all those unacknowledged biases.) Judging other people's decisions then becomes a matter of assessing their reasonableness, relative to one's own ideas of reason. Was A's assault on B justified, given what B did to A's wife? Husbands and wives may have differing views on this point. The judgement is set in the context of competing moral claims, from highly abstract modern ideas about human rights to much older views of human relations. If the answer is no, what is the appropriate punishment for A? If the answer is yes, what arguments support A's justification for his violence? In either case, what signal will this send to the community, and what are the likely consequences for us all? At base, how will this judgement impact on me?

Reasoning uses the framework of agency to explain other people's behaviour, but drains out the blood and colour of strong emotions. Emotions can be involved, but they do not dominate the calculus; to adapt the metaphor, they are not warm enough to melt the crystals of reason. Cool moralizing underlies the pragmatic, sometimes compromised decisions which in principle make law blind to the identity of those it judges, and hence provide the basis of its authority. When emotions do heat up our decision-making changes. Plato's famous analogy from *Phaedrus* of a charioteer trying to control his horses is probably the dominant metaphor, in our richly Greek-imbued culture, alongside that of temperature: we speak of unbridled passions, of reining in our desires, and so on. Modern neuroscience has adapted the ancient image by assigning the role of charioteer to the prefrontal cortex and that of his horses to 'limbic' areas of the brain: those more deeply buried parts which do emotion processing.[27]

HOT MORALS

In practice, emotions come in many forms—not just 'strong' or 'absent'—and reason is less of a muscular charioteer than a teenage novice. Both feeling and reasoning are affected by everything from genes to cultural expectations, from personal experience to what we had for breakfast. What remains of Plato's analogy is that reasoning takes time and energy.[28] When we are tired, or hungry, or need to act quickly, we tend to fall back on less consciously worked-out forms of reasoning, relying more heavily on stereotypes, reacting more quickly and becoming more likely to make errors. We also become more likely to act on our emotions.

Consider a schoolgirl facing the choice of whether or not to talk to a friend who has been ostracized by other members of her gang. She could list the pros and cons of ditching the friend or risking her ingroup membership, assigning each reason a value based on the emotions it arouses: fear of bullying, compassion for her friend, and so on. If it is more important to her to stay within the group, then the friend must go; if the friend matters more, it's time for a display of courage.

If the schoolgirl cares about her friend, however, her reaction to the choice presented by her ingroup is very unlikely to be one of calm assessment. Empathy for their victim, fury and contempt for the bullies, fear for her own position, the blazing conviction that their behaviour is *wrong*—these provide her with a much more colourful, emotional reaction. She is still making judgements about other people's behaviour, but instead of weighing up the pros and cons of that behaviour for her self-interest, with or without considering high moral principles, she is using a much more instinctive framework, in which actions and people are evaluated as good or bad, friendly or hostile, purifying or corrupting, kind or cruel.

Indeed, the term 'evaluated' hardly echoes the rapidity and force of the processes involved. This is 'hot' thought: moral evaluations are swift, automatic and highly resistant to argument.[29] Once given the tag 'abomination', even the most inoffensive targets can find it impossible to gain their judges' favour. Moral stains linger, even when their justifications are explicitly dismissed, and morality involves emotions which can be extremely strong. Tabloid newspapers rely on this emotionality, using language and images which stimulate strong feelings in their readers. Negative emotions are particularly effective.[30] Outrage, fear, disgust, and shock hook buyers.

In Western thinking, as noted above, emotions are traditionally seen as opposed to rational cognition. As we will see in Chapter 5, however, the scientific rediscovery of their importance in decision-making has led to a more nuanced approach. Affective neuroscience regards emotions as memories of the outcomes of our decisions: 'notes to posterity' which saved our forebears having to re-cogitate common situations. If every time you eat a certain food you feel sick, you will come to experience disgust at the sight of the food—even if you do not explicitly remember that the last time you ate it you were sick. The discomfort you feel is a message from your former self: avoid that food! Emotions were already serving this purpose when our species came up with language and symbolic thinking. Now we can use the same mechanisms to write notes to our future selves, storing past evaluations to speed up future reactions. Only in new and complicated situations do we need to work out our choices using reason.

Yet our swift moral judgements are not simply self-centred evaluations ('Is that good for me, bad for me, or neither?'), because they can apply to people and situations far distant from the judge. They can also contradict and even override assessments of personal benefit made using more pragmatic reasoning, forcing us to feel guilt, remorse, and shame. How can this be?

MORALS AND MORAL SYSTEMS

One of the many historically unusual ideas which came to prominence in the scientific and capitalist culture of the modern West was the notion that morality was some kind of optional extra. Some thinkers viewed it as a political tool invented in order to control the masses, a means of oppression which would be superseded by the dawn of better political systems (socialism and communism). Some saw it as a relic of primitive incomprehension rendered increasingly redundant by the advance of science, while some disparaged it as an obstacle to self-gratification.[31] Whether individualists or Marxists, believers in science or social systems or personal expression, these thinkers seemed to assume that morality could be excised from culture, with appropriate re-education.

But morality is more than a cultural predilection. Research in evolutionary and comparative psychology, anthropology, and neuroscience reveals it as a heavyweight player in the symbolic world, with deep roots in human nature. Anthropologists' findings suggest that some form of moral code is universal. All human societies so far examined moralize, sharing basic categories such as obligatory, permitted, or forbidden actions, and children learn the currency of moral judgements as easily as they pick up language.[32] As with languages, the specific content varies, but the structure—the 'grammar'—of moral systems seems fairly consistent across cultures. Just as all languages have nouns and questions, so all moralities have categories like 'pure' and 'impure', 'praiseworthy' and 'blameworthy', 'responsible' and 'not responsible'. Studies in comparative psychology, meanwhile, have found key moral concepts and behaviours—such as fairness, altruism, punishment, and reciprocity—in species other than our own.[33] Noam Chomsky argued that only some innate predisposition could explain children's startling linguistic facility, and a similar position has been made for moral cognition (notably by Marc Hauser's book *Moral Minds*, a thorough summary of recent research on 'animal morality').[34] Whether or not you are prepared to accept the Chomskian position in full, it seems clear that moralizing is not merely a bourgeois fad or system of political control, as radical thinkers have delighted in suggesting. Rather, our tendency to pass judgement on others seems a gift—or curse—of natural selection, like favouring sweet foods or having large brains.

Human beings are thus often described as creatures who live in a moral universe, equipped from birth with an instinctive ability to pick up the moral codes of their society just as they normally learn their ingroup's language. In fact, however, we live in not one moral universe, but at least two. When we talk of 'morality', we often mean the rapid, emotional evaluations of friend and foe, good and evil, which feel as natural to us as breathing. But we may also be referring to moral systems, to abstract ideas of morality, or to rankings of moral feelings derived from scientific sources, all of which are very different from a person's instinctive understanding of right and wrong.[35] Let us therefore distinguish these as 'basic' and 'constructed' morality.

Basic morality is ancient, relatively simple, and similar across many different cultures. As we shall see in the next chapter, it is governed by rules which lead us to look after ourselves and our own, cooperate with ingroup members, prefer the company of people similar to us, be wary of strangers, exploit opportunities to cheat while punishing other cheaters, and react to perceived threats from others with aggressive hostility or even cruelty.[36] Basic morality evolved out of the need to survive as a social animal in a dangerous environment. It survived, ironically, because it was so effective at social control (the Marxists were right to that extent), allowing the formation of highly cohesive groups. By reducing intragroup harm and exploitation through the development and enforcement of social codes of conduct, the capacity to moralize enhanced the genetic fitness of group members.

Constructed morality, as the name suggests, is far more elaborate than basic morality and comes in many culture-dependent flavours. It retains the key features of basic morality: social sensitivity, emotionality, and concern with what *should* be done, rather than what happens to be the case. But its reach is wider, encompassing principles formalized in bodies of law and political theory—including, for example, demands for animal welfare and animal rights, just war theory, and the doctrine of universal human rights. It developed out of basic morality in response to the pressures of cultural innovation, as human social systems became more complex, institutionalized, and abstracted. It can vary considerably even within a given society at one time (e.g. between different ethnic groups) or across short periods of time (e.g. if war breaks out).

Constructed morality, like reasoning, is a product of leisure; it only became possible once human beings had gained enough control of their world to give themselves some time to think. As such, it gains its reach from the power of cool reason to abstract across the whole human race, or even beyond. Yet unless it can link its high abstractions to the fierce energies stirred up by moral emotions it will have little effect on actual human responses.[37] In practice, therefore, this is what moral systems do.

A society's constructed morality may be more, or less, consistent with basic morality. Much of our puzzlement about our moral lapses comes from tensions between the two. In the universe of basic morality, for example, some cruelty is acceptable—when it is directed at threatening outsiders. Moral protection covers the ingroup. It may be extended, in part and provisionally, to those outgroups with which the ingroup trades or intermarries, but it does not apply to enemies (and a trading partner can quickly become an enemy).[38] Cruelty can benefit the ingroup, not least by terrorizing or even eliminating the competition. Within a group, moral codes can also vary. The constructed moral systems of specialized warriors, for instance, have much more in common with basic morality than those of the civilians they protect. Many terrorists' morals (yes, they do have them) also look like basic moralizing, however dressed up in haughty ideology.

Many elaborations of constructed morality, by contrast, imply or state directly that cruelty is unacceptable in any circumstances. Just war theory depends on the idea that aggression is justified and proportionate; the doctrine of universal human rights effectively proclaims the entire human race an ingroup, and the animal rights movement insists that cruelty is unacceptable even to other species.[39] Nor are attempts to extend a ban on cruelty beyond the ingroup restricted to the modern West (Buddhism and Jainism are two examples of other traditions which take the idea extremely seriously). It is also worth noting that even in cultures which proclaim their civilized values—like fifth century BC Athens and the modern United States—basic morality all too often trumps the constructed kind, with the powerful (mostly or exclusively male, rich, and white) elites favouring their members and exploiting the rest. Both societies, after all, were founded on slavery, discrimination against women and ethnic minorities, and the restriction of democracy to exclude much of the population.[40]

The United States, faced with very different pressures, has adapted as Athens did not, promoting equality and democratic values to a historically extraordinary degree (women, for example, did not hold political office in Athens). Yet all too often—in international politics, for instance—the rules are still applied selectively to Americans but not their putative enemies, as the moral disasters surrounding the 'war on terror' have shown.[41] This favouritism is entirely consistent with basic morality but does not sit well with constructed moral systems which claim to be universally applicable, as the US model does. Constructed moral beliefs can be passionately held if they resonate with more basic moral emotions; but where the two conflict the older system is likely to win out.

THE EMPATHY SWITCH

To illustrate the distinctions between basic and constructed morality, consider the following scenario. A girl (let us call her Jane) is watching the news when a report comes up that a young woman has died. The victim is named as the daughter of a public figure whose arrogant mistreatment of others Jane detests, despite never having met him. Jane feels a mixture of emotions: lukewarm sympathy, weary recognition of evil done, perhaps relief that the death has hurt no one she knows, maybe a touch of *Schadenfreude*, and the inevitable startled thrill of being connected to the event, if only through having heard of the celebrity father. That is her cool reasoning response, which evaluates the event with respect to Jane's self-interest and her more abstract moral values, estimating the costs and benefits to her and presenting the result in mildly emotional terms.

Although she dislikes the bereaved parent, she has no strong personal connection with the tragedy. Cold-blooded pragmatism might thus suffice—but Jane is a moral young woman, and immediately feels guilty for being so callous. That is the influence of constructed morality; people in Jane's culture, particularly women, are expected to be caring, and she knows it. She may therefore bolster her sense of sympathy, squash the *Schadenfreude*, reprove herself for the thrill and her compassion fatigue, perhaps even consciously articulate the thought that even the arrogant do not deserve to lose their daughters.

Then the newsreader moves on to give more details. The girl, on the evening of her twenty-first birthday, was raped, stabbed repeatedly, and left to die. Jane involuntarily shudders. The pity and horror she feels is her basic moral response. It leads her to deeper sympathy for the family and an urgent desire to see the attacker punished. She now feels connected to the girl in a way she did not before.

All Jane's responses use emotions, albeit of different intensities. And in one sense the meaning of the event has hardly changed: the girl is just as dead, no matter how she was killed. What did change was how Jane responded to the news. She began to empathize. Empathy, it seems, may serve as a mental switch, engaging our basic morality in cases to which we feel connected; cases which have implications for us. As we shall see, empathy is not the only switch connecting us to others. Beliefs can also trigger moral responses—or suppress them.

Summary and conclusions

Cruelty is as old as humankind, if not older. At its core lies unjusti-fied voluntary behaviour which causes foreseeable suffering to an unde-serving victim or victims. It may involve physical aggression, or more subtle abuse, but its aim is to make its targets suffer physically or psychologically.

Cruel behaviour is also voluntary. Though the urge to hurt can become so strong that it may be called compulsive, it is usually still controllable to some extent. Not always, hence the need for terms like 'running amok', 'losing it', 'going berserk'. But that pattern is quite different from so-called compulsive criminals like rapists, paedophiles, and serial killers. They plan their crimes with care, sometimes desisting at the last minute, sometimes suppressing the urge to kill for months or years. We judge both types of damage to be inflicted by responsible agents, but we take account of mitigating factors with moral and legal flexibility. Free agency is a matter of degree.

Cruelty is, moreover, a moral concept. People who harm others must justify their actions; cruel behaviour is liable to be punished by those who find the justifications lacking. Depending on the circumstances, we make those moral judgements in various ways: swiftly, using the moral emotions inherited through natural selection, or more slowly, using self-interested reasoning and the moral principles provided by our culture.

If constructed morality approves the justification of someone who does harm, we need not call them cruel. Though our basic moral instincts may tell a different story, we can learn to control them when it is politic to do so—that is, when self-interest suggests a temporary gag on self expression. If self-interest, social norms, and ancient instincts are all in register, however, moral judgement can be a devastating force, capable of whipping up a lynch mob. Cruelty brings down particular opprobrium. Yet cruelty is not a monolithic concept. Callousness, that mark of Cain, is despicable, but still in principle redeemable. Sadism goes beyond Cain, deep into the realms of evil. As we shall see, however, moral passions often interpret callousness as sadism—and calling someone sadistic leaves them open to abuse. Moral judgements, paradoxically, have been used to justify extraordinary cruelty.

Moral instincts may be part of the human recipe, but moral judgements about cruelty are shaped by much more than evolution. In the next chapter we investigate how much more, by asking who decides what counts as cruelty.

Chapter 2

Quis judicat? Who decides?

Given the chance I'd hang the lot.

(Benjamin Britten and Myfanwy Piper, *Owen Wingrave*)

Healthy human beings develop both rational and emotional methods of evaluating other people's behaviour. Our judgements are flavoured with a mixture of the two. Sometimes cool, pragmatic reasoning dominates—with or without reference to moral principles. Sometimes basic morality heats up the calculations. As we have seen, empathy may be one of the cues which triggers basic moralizing. This is not to say, however, that the same stimuli will always evoke the same response. Personal history matters. In the last chapter's examples of the cat and the old lady, my description of the likely responses might not be yours. If you detest cats, or think, as many people used to, that they are unfeeling machines, you may use reasoning to judge that example. If an old lady you care about was recently mugged, you may react with outrage to my suggestion that rationality dominates. If neither cats nor old ladies get your moral juices flowing, some things will. Confronted with a story about, say, a man who rapes a toddler, even veteran criminals may shiver with revulsion—and act on their loathing; in prison, sex offenders are at high risk of attack from other inmates.[1] The intensity of our reaction, however, will depend on our culture, our personal beliefs, the social roles we play, and so on. An experienced (read: desensitized) child pornographer is likely to be less affected than a mother, to state the obvious. And children, even toddlers, do get raped. Clearly one person's moral poison is another one's meat.

Context also matters. People placed into specific situations, or playing specific roles, can make very different moral judgements from what they would consider their 'normal' response. A notorious example of this situation-dependence is the well-known Stanford Prison Experiment, in which Philip Zimbardo and colleagues placed American students in the roles of guards or prisoners, resulting in behaviour by the guards which degenerated so rapidly into serious physical abuse that the experiment had to be stopped early.[2] The student guards were not sadists, but their understanding of prison guards' roles included the grasping of power without responsibility (the latter had been displaced to the experimenters). Given that power, they used it against the prisoners they saw as rebellious and potentially dangerous. The same story of escalating abuse has been played out repeatedly in poorly managed prisons, before and since Zimbardo and colleagues' research.

The greatest difference in roles, with respect to cruelty, is that between perpetrator and victim. The word 'cruel' is typically used not by perpetrators themselves but by victims or third parties, to condemn the behaviour and criticize the morally culpable perpetrator. The latter, of course, often sees things very differently. Victims interpret the perpetrator's actions as cruel and deserving of moral condemnation, while perpetrators often view their impact on the victim as an unfortunate necessity, a side-effect, a benefit, or an irrelevance.[3]

Quis judicat?

Quis judicat? Who is to decide? Who but that reluctant behemoth, society. No perpetrator acts, no victim suffers, in total isolation, even though they may kill, or die, alone. Both are embedded in a cultural background, and usually in one or more social networks, groups full of instant moralizers. For perpetrators, these groups may present the threat of legal punishment or vengeance from allies of the victim. There may also be other consequences for cruel behaviour. Group members can and do show their disapproval with otherization techniques like negative stereotyping and ostracism ('he's a bad sort, best avoid him'). In the small groups in which our ancestors are thought to have lived, withdrawal of social support could be lethal, providing strong disincentives to further cruelty and motives for future good behaviour.

Today, in the developed world at least, social isolation is less immediately lethal, since vital resources like food, housing, and heat tend to be supplied irrespective of one's social standing. The loneliest people in Britain can still do their weekly shopping at a supermarket; the wickedest Americans, held in solitary confinement, still get fed. Nevertheless, research suggests that social isolation is extremely bad for us, significantly raising the likelihood of disease and early death.[4] We are also highly sensitive to cues suggesting social disapproval. This is why fashion-driven consumerism, a technique for converting heartache to profit without curing the heartache, is so adept at prising money from us.[5] Yet the same propensity helps to keep cruelty's worst excesses comparatively rare. The thought of all that condemnation, either from others or one's own rebuking conscience, makes viciousness unappetizing.

Both perpetrators and victims (or their allies and dependants) seem to conceive of their actions in relation to a third-party audience—someone who may find the evidence perpetrators try to hide, or give the victims the justice they cry out for. In cases where the crime itself is unobserved, conceptions of the audience may vary. It may be 'society' in the sense of the person's fellow-citizens, or, in cases of state-organized killings, other nations or supranational bodies like the United Nations. It may be a very small ingroup, like a soldier's closest colleagues, or it may be an ideal of the human race in general, or future historians, or God. However conceptualized, both perpetrators and victims view this observing Other as a judge they may have to persuade. Victims are more likely to be rescued, compensated, or avenged if those who judge the perpetrators are convinced that the victims suffered unfairly and the perpetrators acted freely. Perpetrators have the opposite task, but they can be just as aware of their audience, especially in advance, when they are persuading themselves and others to condone or even join in their cruelty. In a speech to his army chiefs just before the outbreak of the Second World War, Adolf Hitler used examples of previous mass killings (the medieval depredations of Genghis Khan and the Turkish slaughter of Armenians in 1915) to argue that the Wehrmacht could act with impunity: 'Our strength consists in our speed and in our brutality. Genghis Khan led millions of women and children to slaughter—with premeditation and a happy heart. History sees in him solely the founder of a state.... Who, after all, speaks to-day of the annihilation of the Armenians?'[6]

The working definition of cruelty adopted earlier viewed it as unjustified voluntary behaviour which causes foreseeable suffering to an undeserving victim or victims. Justification can be considered in highly abstract terms, but it also has its passionate side, tapping into our strong sense of what is fair (or otherwise).[7] As a moral concept, cruelty is embedded in a framework which provides human beings with finely honed capacities for detecting unfair treatment and assessing justifications for behaviour. If the victim(s) deserved punishment, then harm-doing action against them was fair, not cruel. If the action was necessary (in that the perpetrator had no alternative), involuntary (the perpetrator was forced to act), or justified (in that the perpetrator had acceptable reasons for acting as he or she did), then the action likewise lacks the moral stigma associated with the label 'cruelty', unfair on the victim though it may be. Some situations simply cannot be resolved without someone getting hurt.[8]

Unsurprisingly, victims—or their advocates, who may of course have their own agendas—frequently emphasize their innocence and suffering, attack the unjustifiable motives of the perpetrators, and make much of the latter's freedom and agency. Situational explanations from social psychologists and historians, which de-emphasize human agency and favour causes in the person's environment, like peer pressure or obedience to authority, are often not well received by those fired-up with moral outrage.[9] Loading opprobrium onto an environment is hardly psychologically satisfying, especially if done in the cool language of scientific reasoning. Instead, victims and their allies may fuel hot moral emotions by describing the harm-doing in emotive terms, as Daniel Goldhagen did in *Hitler's Willing Executioners*, emphasizing barbarity and degradation in an attempt to evoke empathy from observers.[10] Goldhagen's book may be the work on Nazi Germany most likely to make academic historians of that period roll their eyes in despair; it was also a media sensation. Moral outrage engages us in a way that academic pragmatism cannot always match.

Perpetrator pragmatics

... when evil is done the whole art of oratory is a screen for it.

(Thucydides, *History of the Peloponnesian War*)

Perpetrators, like victims, may aim for empathy by presenting themselves as reasonable and likeable people; but empathy alone may not suffice to get them off the moral hook. To evade the charge of cruelty they must use other strategies. One approach, still exemplified by many Turks vis-à-vis their treatment of the Armenians, is simply to deny that massacres happened, or that they involved anything more than run-of-the-mill atrocities of war (regrettable, but never genocidal).[11] As Turkey is increasingly finding, however, it can be stressful to maintain denial against the challenges of outraged third parties (and indeed Armenians themselves). This is still more the case in those comparatively few situations when judgement rapidly follows the crime. Given time, perpetrators can obfuscate, lie, blur memories, and perhaps convince even themselves that they are innocent. (Meanwhile those who might judge them forget their crimes.) Swift condemnation makes denial harder, placing extra pressure on those who know perfectly well that their behaviour created the corpses in question.

Nowadays, downplaying the victim's suffering is likewise problematic. Under the stern gaze of modern technology, bodies can bear witness to torture years after being dumped in mass graves. Media spotlights can blaze on victims' agonies and leave perpetrators blinking in the glare. Otherizing ideologies, with their intense dehumanization of victims, are not always sufficiently protective; a dead child can grab the heartstrings of even the most indoctrinated killer. Besides, in this age of global human rights and the fellowship of man, woman, and child, justifications which blatantly dehumanize—using terms like 'subhuman', or comparing victims to diseases—tend to be frowned on. Where, then, to turn for the perpetrator in search of a justification?

One favourite strategy, especially given that many perpetrators are highly egocentric characters, is to focus not on the victim, but on themselves.[12] For instance, instead of downplaying their victims' suffering they can downplay their own free agency, claiming that their actions were necessitated. The necessity defence comes in a variety of flavours, one of the more notorious being the 'following orders' argument aired—and rejected—when the Nazi regime was put on trial at Nuremberg. On a smaller scale, criminals and their defence lawyers have not been slow to use science (admittedly usually media-bastardized versions), seizing on factors which they hope will explain away their agency and responsibility. From genes for aggression to under-par pre-frontal cortices, from environmental pollutants to the notorious (and mythical) 'Twinkie' defence', it's amazing how feeble and passive people can be when it comes to doing harm.[13] They become leaves blown about by the winds of circumstance, reacting to their situations like machines to their inputs. It is odd how, when life and liberty are not under judicial review, the same people act as if they are agents, enjoying the 'illusion' of free will and behaving as if they have some say in the matter of their lives. As Shakespeare has Edmund remark in *King Lear*:

> This is the excellent foppery of the world, that when we are sick in fortune, often the surfeits of our own behaviour, we make guilty of our disasters the sun, the moon and the stars; as if we were villains on necessity, fools by heavenly compulsion, knaves, thieves and treachers by spherical predominance, drunkards, liars and adulterers by an enforced obedience of planetary influence; and all that we are evil in by a divine thrusting on. An admirable evasion of whoremaster man, to lay his goatish disposition on the charge of a star. (I. ii. 118–28).[14]

This is what philosophers call 'hard' determinism—the idea that every action is bound tight into a causal chain, leaving no room for us to exert free choice. Philosophers, however, reach for the universal at every opportunity: if hard determinism applies at all, it presumably applies as much to doing the shopping as to killing your wife or beating up your workmate. Using it selectively is not allowed, yet this is what perpetrators regularly do.

For those perpetrators who cannot make chemicals or brain abnormalities sound convincing (for example, when an entire community of 'normal' people is involved), another way to boost one's claims of necessity is to claim self-defence. Self-defence arguments are common, especially when civil wars, insurgencies, or political revolutions are involved, as they frequently are.[15] 'Kill or be killed' may carry conviction in a faction-riven society like Rwanda, where there was a genuine threat of invasion and massacre (indeed, both occurred sporadically for years before the 1994 genocide, killing thousands of people). Self-defence arguments are harder to maintain when the target of the hysteria is a small, unthreatening, generally law-abiding minority, as Germany's Jews were prior to the Holocaust. Perpetrators therefore need to stretch their logic to encompass threats the rest of us would never even notice, so that tiny minorities can become the seeds of cultural and/or biological doom, or the fifth column of an external enemy, or both.[16] Indeed, when committing mass killings all perpetrators are likely to need justifications of this sort (if they are ever asked to justify their actions, which most are not), because mass killings tend to sweep up the clearly harmless— elderly people, the disabled, pregnant women, and babies—as well as the potentially dangerous.

The answer, unsurprisingly, is to reclassify everyone as dangerous, even those who pose no conceivable threat of physical attack. Oskar Groening, an SS man, was sent to work at Auschwitz in 1942, when he was 21. Asked, much later, why children were killed in the gas chambers, he answered: 'The children are not the enemy at the moment. The enemy is the blood in them. The enemy is their growing up to become a Jew who could be dangerous. And because of that the children were also affected' (note how the euphemism plays down the children's suffering).[17] One need not, however, be a Nazi to proffer such rationalizations. During the First World War, as the Turkish government attacked its Armenian minority, its Minister of the Interior is said to have told a German reporter: 'We have been reproached for making the distinction between the innocent Armenians and the guilty; but that was utterly impossible, in view of the fact that those who were innocent to-day might be guilty to-morrow.' Compare the El Salvadoran captain telling his troops to murder children in El Mozote, 1981: 'If we don't kill them now, they'll just grow up to be guerrillas.' Three different cultures, one lethal rationale.[18]

This is where the essence trap described in the Introduction can be so devastating, because by relocating a person's 'dangerousness' from their behaviour to their nature, it cuts off the possibility of changing them. Whether applied to perpetrators or victims, otherization homogenizes people into crowds, blurring their differences. Otherizing metaphors like 'swarm' and 'plague' stick much more easily to a large and faceless crowd than to a set of distinctive individuals.[19] Yet this is as misleading for perpetrators as it is for victims, since both include varied subsets of humanity. Victims can be cruel, arrogant, deceitful; they can even be perpetrators. Similarly, a single group of people committing a war crime may contain ideologically motivated fanatics, greedy scavengers, psychopaths out to avenge some minor offence, technicians proud of their skill in getting the job done, junior members longing to impress their superiors, and novices trying not to faint, cry, or throw up. Calling all of these perpetrators evil makes no sense, just as calling their victims dirty or treacherous makes no sense.

When it comes to otherizing victims the language of sickness and waste is particularly effective. Nazi alchemy, for instance, transmuted the 'Jewish question' from selfish persecution to a life-or-death war against disease, using the justification of hygienic necessity to legitimize cruelty on a massive scale. Unlike all other alchemical attempts, but like every atrocity before and since, Hitler's state turned something valuable—irreplaceable, individual human lives—into something utterly devalued—nameless corpses. The transmutation of low-value, 'base' material into gold was the medieval alchemist's dream, fabled but never achieved. In human terms it is the dream of redemption; and it has been known to come true: people do turn their lives around, change their ways, forbear from viciousness, gradually learn to love. But redemption, like any entropy-defying process, takes energy. The reverse transmutation, from person to human waste, takes much less effort; and that is what happened at Auschwitz, Tuol Sleng, and all their hideous kin.[20] More than the physical agony and deprivation, it is the stripping down and degrading of human meaning which makes such cruelty so terrible.

LIKING AND BIAS

Death camps and genocide, one would have thought, are unequivocal; yet there are people who deny the Holocaust, and not all of them from a position of ignorance. They do so because their beliefs make acceptance impossible (later we shall examine how such biases exert their blinding effects). In many less overwhelming cases of cruelty, the observers whose role it is to judge may likewise be susceptible to various more or less widely acknowledged biases.[21] For example, a high-status or likeable perpetrator is less likely to be judged cruel, *ceteris paribus*, than is a member of a group distinct from and disliked by the observer (an outgroup member). Similarly, observers are more likely to turn a blind eye to actions against a lower-status victim (especially a member of what John Conroy terms 'the torturable class' of society's marginal figures) than one who more closely resembles them in status, appearance, and/or behaviour.[22]

A key determinant, as noted earlier, is the emotional connection between people. The more one likes or cares for another person, the easier it is to empathize with their feelings. But empathy need not be the final word. Other emotions, like happiness or sadness, can modify its intensity; disgust can often suppress it altogether. Cognition and learning are also relevant: what one believes about a person has a huge effect on whether or not one empathizes with them. Moreover, empathy for others can work both ways, especially when it comes to cruelty. Applied to victims, it makes us righteously moral, defenders of their shattered human rights. Applied to perpetrators, however, it can leave us uneasily shutting our eyes to abuse and falling back on the cooler language of reason. Moral frameworks may gain force from strong emotions, but they simultaneously lose consistency. We cannot care passionately for everyone.

Basic morality is local morality. It lacks the 'all men are equal' absolutism of constructed moral systems, so treating some men as rather more equal than others is less of a problem. Whether we judge an action as cruel depends on whether we see it as gratuitous—not justified by the agent's goals—and that in turn depends on who the agent is and their relationship with us.[23] What we judge their goals to be, and how valid and important we think they are, will vary depending on our priorities, as well as our opinion of the agent.

A typical adult human is born with an instinctive capacity for empathy. Appropriately nurtured, that capacity will leave them flinching and distressed when confronted with evidence of cruelty. Our moral systems rapidly respond to signs of fear, distress, and pain in other people.[24] Bias sets in after the initial reactions, in situations when the perpetrators are not distant Others but liked or respected people, friends, or even just members of a salient ingroup.

Which ingroups are salient depends on the situation. If Moshe is an Israeli, or supporter of Israel, who sees footage of Israeli soldiers beating Palestinian children, his salient ingroup is likely to be 'Israeli', but how salient it is will vary with the attention he gives the footage (and what else is happening at the time), his commitment to Israeli ideology (the extent to which those beliefs are important to his sense of self), and so on. The more salient the ingroup, the greater the internal conflict when Moshe sees people like him behaving cruelly. How is he to resolve that conflict? One option is to simply disbelieve the reality his own eyes show him, but that way lies psychosis. Another is to reinterpret it, for instance by deciding that the perpetrators are not, after all, 'true Israelis'; yet this is not easy when the soldiers are clearly marked as state representatives. Or he could reject his self-identification as an Israeli, but that may be too painful even to contemplate. And so Moshe, like innumerable others before him, will stumble down the slippery path of excusing what at first sight seems inexcusable, denying that what looks like cruelty actually *is* cruel by finding justifications for the soldiers' behaviour. Justified actions imply no hunger to hurt. This search for reasons is by far the easiest of the options available to Moshe. Less committed observers who condemn him for condoning an atrocity may fail to understand just how much he would lose by making any other choice, caught as he is between the Scylla of doubting reality and the Charybdis of amputating a massive chunk of his own identity.

The problems of classification caused by conflicting ideological biases ('he tortures, you are excessively harsh, I am driven by unfortunate necessity') do not, however, mean that cruelty is entirely in the eye of the beholder. Even the most partisan observer is likely to recoil at some atrocities. Americans who had vigorously supported a war in Iraq, in the full knowledge that war means killing people, were horrified by the images from Abu Ghraib, in which Iraqi prisoners were tortured and humiliated. What distinguished Abu Ghraib was the gratuitous nature of the victims' suffering, which seemed pointless except as entertainment for their captors. In other words, some actions inflict suffering which simply cannot be justified by the agent's goals, however favourably viewed. Many wartime atrocities are of this kind, as, for instance, when a soldier who is ordered to kill opponents tortures them first. The additional pain inflicted is gratuitous, in that it is not required as part of the killing process; there are other, easier, quicker ways to kill. Indeed, by delaying the soldier and thus reducing the overall kill-rate, cruelty may at times be counter-productive. Excessive cruelty hints at morally abhorrent desires, at enjoyment of the pain inflicted for its own sake rather than as a means to some other end. It is this delight that epitomises the worst extremes of cruelty: behaviour which causes suffering to an undeserving victim or victims for no other reason than to please the perpetrator.

In the nineteenth century the English language acquired a new name for this repellent hunger: sadism.

The hunger to hurt

There were people whose participation awakened in them the most evil sadistic impulses. For example, the head of one firing-squad made several hundred Jews of all ages, male and female, strip naked and run through a field into a wood. He then had them mown down with machine-gun fire. He even photographed the whole proceedings.

(SS-Obersturmführer Albert Hartl, speaking about the Holocaust in 1957)

We owe the term 'sadism' to the European Enlightenment, and specifically to the writings of Donatien Alphonse François, Marquis de Sade (1740–1814), who in works such as *Justine* and *120 Days of Sodom* linked cruelty to carnality and so gave sadism its sexual flavour. Sade also spells out the connection between cruelty and sexual satisfaction: selfishness. In *Justine*, for example, the libertine monk Clement argues that sharing sexual pleasure with a woman diverts a man's resources towards her, and away from his aim of maximal enjoyment:

> And what then is this necessity, I ask, that a woman enjoy herself while we are enjoying ourselves? in this arrangement is there any sentiment but pride which may be flattered? and does one not savor this proud feeling in a far more piquant manner when, on the contrary, one harshly constrains this woman to abandon her quest for pleasure and to devote herself to making you alone feel it? Does not tyranny flatter the pride in a far more lively way than does beneficence? In one word, is not he who imposes much more surely the master than he who shares? ... To love and to enjoy are two very different things.[25]

Hurting people who are weaker than they are is so deeply pleasurable for Sade's anti-heroes that they pursue it with complete disdain for legal, moral, or humanitarian constraints. Cruelty 'flatters their pride', giving them a sense of control; might is right in this predatory world. They revel in dreaming up new variants on the themes of pain and distress: violent rape and incest, paedophilia, torture, even vivisection (of children, not animals). Sade, however, is not only interested in describing cruelty as inherently rewarding (sadism). He also argues for a nihilistic vision in which indifference to suffering is a fact of nature and compassion a learned and debilitating convention. In doing so he explicitly emphasizes callousness, the imperative for the Sadeian protagonists to disregard the victim as a person:

> The necessity mutually to render one another happy cannot legitimately exist save between two persons equally furnished with the capacity to do one another hurt and, consequently, between two persons of commensurate strength: such an association can never come

into being unless a contract is immediately formed between these two persons, which obligates each to employ against the other no kind of force but what will not be injurious to either.

So far, so good:

> ... but this ridiculous convention assuredly can never obtain between two persons one of whom is strong and the other weak. What entitles the latter to require the former to treat kindly with him? and what sort of a fool would the stronger have to be in order to subscribe to such an agreement? I can agree not to employ force against him whose own strength makes him to be feared; but what could motivate me to moderate the effects of my strength upon the being Nature subordinates to me?[26]

This is a psychopath's charter: To the strong: I'll be civilized. To anyone else: I'll do what I like, and fuck you.

Sade's brutal perpetrators (most but not all male) clearly illustrate how the two faces of cruelty—the indifference required to be cruel, and the delight which drives its worst excesses—intermesh to fuel increasing violence. Indifference is essential: otherwise empathy would cause too much distress, deterring action. But indifference to suffering leads to callousness: harm done in the pursuit of other goals. When those goals themselves *become* the creation of pain we reach the extremities of cruelty for its own sake: the evil we call sadism. That term, despite its etymology, need not apply exclusively to sexual abuse: there are other rewards than orgasm. I will use it in the broader sense to refer to the cruelty which occurs when inflicting suffering becomes associated with any type of reward: sexual, chemical, financial, social, and so on.

Sade's torturers, like many since, do not lack what psychologists refer to as 'theory of mind', or the ability to mentalize.[27] They are adept at getting inside their victims' heads—several take the time to hold lengthy, interested debates with *Justine*'s heroine, in between bouts of abuse—and the text offers no evidence that they are deficient in the capacities to reason, imitate, experience pleasure, feel pain, or recognize the anguish of their victims. What they do appear to lack is any motivation whatsoever towards alleviating that anguish. (Justine herself, by contrast, clearly feels fondness, sorrow, and compassion for other victims, identifying with them as they undergo their sufferings.) These perpetrators are neither aliens nor automata; they recognize not only pain but empathy. (Why, some of them find these symptoms quite amusing.)

Callousness is one thing. We may sympathize or condemn, but we can force ourselves to understand. We may even shiver and think 'there, but for the grace of God, go I'. But fewer of us care to approach the idea of killing someone just because we feel like it. Sadism is unacceptable, inhuman; a step beyond the framework of necessity into the abyss of moral revulsion. At least, that is how most of us like to see it, most of the time. It can also be deeply exciting, though that possibility is not easy to accept. (We will return to the pleasures of sadism in Chapter 8.)

Callous perpetrators may not understand the intensity of the agonies they inflict, or may not care. Indifference to suffering, however, is not equivalent to enjoyment of it. Only when suffering itself becomes the primary goal do we move from callousness to sadism. Sadistic cruelty occurs for its own sake, rather than as a means to other ends. It is the most feared, most condemned, and least understood type of cruel behaviour: overestimated, overglamourized, and overloaded with moral and psychological baggage; perhaps the human behaviour most likely to make us tumble headlong into the essence trap of seeing sadists as evil, monstrous villains. Human weakness might lead us into callousness, but the leader who coaxes or drags us into the darkness of sadistic cruelty, the Hitler or Pol Pot or Stalin or Genghis Khan, is the evil genius who bears the heaviest moral burden. And there must always be someone onto whom we can dump that burden, because sadism is evil, and evil is Other, not Us.

The distinction between sadism and callousness reflects my earlier distinction between basic moral and more consciously constructed ways of judging people. Both dichotomies contrast 'cold-blooded' practicality—which, if anything, pays too little attention to the emotions—with 'heated' reactions which seem to depend upon them. When we empathize, or when strongly negative emotions drive us to condemn, we pay less attention to the goals and costs and benefits which we might otherwise have rationally assessed. We loathe and despise callous perpetrators, but we still understand them to be human: they will respond to incentives and punishments; in this they are reasonable, like us. Sadists, on the other hand, are fearsome in the way a tiger or deity is fearsome. They are beings with strong and extremely threatening desires, impervious to the reasons and incentives which would otherwise give us a means of changing their behaviour. In practice, of course, the dichotomy is not clear-cut. We can swing between seeing a perpetrator as evil, irredeemable, and impenetrably sadistic and seeing him or her in more human terms, acting and failing for reasons we can recognize.

Callousness and sadism in practice: an example

Callous behaviour deliberately aims at harming the victim or victims. The suffering may be downplayed, either by using otherization to downgrade the victims' perceived capacity to feel or by simply choosing not to think about their anguish—but some miseries are so obvious that denial, at least for those closest to the action, becomes an immense strain. Such explicit suffering may not, however, imply sadism on the perpetrator's part. He or she may see the victims' trauma as a nasty but necessary proximal goal, a step towards achieving a higher and worthier aim. If belief in that aim is strong enough, it can serve to justify astonishing atrocities, at least in the minds of those who commit them. Even Nazis closely involved with the day-to-day running of the Holocaust retained the ability to distinguish between callousness and sadism. Becoming a perpetrator does not necessarily require one to jettison one's morals wholesale, *au Marquis*. What it does require, however, is callousness. For the Nazis, disregard for victims was an integral part of the warrior's code.

In October 1943 Heinrich Himmler made a speech to SS members in Posen (now Poznań in Poland) which has become notorious for its praise of callousness.[28] In this speech the Reichsführer-SS notes how 'duty' overrode aversion when the task was to 'stand comrades who had failed against the wall and shoot them' (a reference to the Nazi Party's purge of Ernst Röhm and his supporters in 1934). He acknowledges the emotional stresses involved—'everyone shuddered'—but adds that they are overcome: 'everyone was clear that the next time, he would do the same thing again, if it were commanded and necessary'. Himmler goes on:

> It is one of those things that is easily said. 'The Jewish people is being exterminated,' every Party member will tell you, 'perfectly clear, it's part of our plans, we're eliminating the Jews, exterminating them, ha!, a small matter.'
>
> And then along they all come, all the 80 million upright Germans, and each one has his decent Jew. They say: all the others are swine, but here is a first-class Jew.
>
> And none of them has seen it, has endured it. Most of you will know what it means when 100 bodies lie together, when there are 500, or when there are 1000. And to have seen this through, and—with the exception of human weaknesses—to have remained decent, has made us hard and is a page of glory never mentioned and never to be mentioned.

Decency—allowing for the human weaknesses which tend to surface during the process of producing lots of corpses—is entirely compatible with callous elimination of the enemy. Himmler's argument is clear. 'We have the moral right, we had the duty to our people to do it, to kill this people who wanted to kill us . . . We have carried out this most difficult task for the love of our people. And we have taken on no defect within us, in our soul, or in our character.'

Himmler's message had already reached the lower ranks. SS-Obersturmführer Karl Kretschmer, writing to his wife and children in 1942, writes that: 'we are fighting this war for the survival or non-survival of our people. . . . We have got to appear to be tough here or else we will lose the war. There is no room for pity of any kind. . . . Only a person who has himself firmly under control can judge or rule over others . . . it is a weakness not to be able to stand the sight of dead people; the best way of overcoming it is to do it more often. Then it becomes a habit.'[29]

Indecency is another matter entirely. Himmler's example at Posen was SS men who stole from the corpses; they were to be executed without mercy. On other occasions he lumped together self-seeking motives, sex, and sadism as criminal behaviour (all three are characteristic of 'the Jew', as stereotyped by Nazi leaders). Again, the Reichsführer's justification was ideological: 'But we do not have the right to enrich ourselves with even one fur, with one Mark, with one cigarette, with one watch, with anything. That we do not have. Because at the end of this, we don't want, because we exterminated the bacillus, to become sick and die from the same bacillus.'

Sadism, like looting, was a perversion of heroic conduct. Described as pathological and bestial, it was a matter of great concern to some senior Nazis. Johannes Blaskowitz, Eastern Territories Commander in 1940, complained: 'If high-ranking SS and police officials demand and openly praise acts of violence and brutality, before long people who commit acts of violence will predominate alone. It is surprising how quickly such people join forces with those of weak character in order, as is currently happening in Poland, to give rein to their bestial and pathological instincts.' Another officer commented as follows on the operation to eliminate all a town's Jews: 'As far as the manner in which the action was carried out, it is with deepest regret that I have to state that this bordered on the sadistic.' A year later, a decision from the Reichsführer-SS on 'how to punish men for shooting Jews' distinguished such actions on the basis of motive. Politically motivated killings were not to be punished 'unless punishment is necessary for the purpose of maintaining order'. Sadism, by contrast, was fully culpable: 'Men acting out of self-seeking, sadistic or sexual motives should be punished by a court of law and, where applicable, on charges of murder or manslaughter.'[30]

The Nazis saw themselves as engaged in a fight for their survival, a total war against a powerful international 'Jewish-Bolshevik' conspiracy (merging Russians and Jews into a single social organism, as if Jews in Russia had never been persecuted). In total war, as Hitler told his generals, the goal lay not 'in reaching certain lines, but in the physical destruction of the enemy'.[31] There is evidence that the Nazis perceived the motives of those they were fighting in similarly totalist terms. One, in a letter home, writes: 'The bomb attacks have, however, shown what the enemy has in store for us if he has enough power. You are aware of it everywhere you go along the front. My comrades are literally fighting for the existence of our people. The enemy would do the same.' Another, in post-war testimony, recalls his comment after watching a massacre in Lithuania: 'May God grant us victory because if they get their revenge, we're in for a hard time.' Given such beliefs, callous elimination of enemy personnel was, from the Nazi viewpoint, fully justified. Sadism, in principle at least, was not; in practice we know that some Nazis, though not all, devised ingenious cruelties for their victims. Yet the very fact that Himmler and other commanders acknowledged and deplored the existence of sadism suggests that they recognized its importance. These were not men without morals. To judge otherwise is to fall into the essence trap: comforting, maybe, but mistaken.[32]

Summary and conclusions

Killing in self-defence, in a fight for one's very existence, is widely seen as morally justifiable. It is thus, strictly speaking, not cruel, because the victim is also an instigating attacker and would-be killer, and hence not undeserving. Killing for pragmatic reasons is more culpable and more likely to be seen as cruel, but it is at least comprehensible. To enjoy the act of killing or hurting, however, is generally considered morally repugnant: in the ranks of evils sadism trumps callousness. The more we perceive sadistic motives to be relevant in an act of human harm-doing, the more likely we are not only to condemn it on moral grounds, but to do so with powerful moral feelings of outrage and disgust for the perpetrators.[33]

Confronted with an atrocity, we may well react with horror—although the intensity of our feelings will depend on whether we are stressed, distracted, on guard against 'weakness', or atrocity-saturated.[34] Assuming we have not been numbed to callousness, our initial horror will have much to do with exactly how we are confronted. Seeing a dead body on the news is very different from tripping over one; and both have more impact than hearing or reading reports of numbers killed. The physical shock and revulsion caused by death and damage reflects the presence of the cues we have evolved to treat as warnings: blood and entrails, mutilated or decomposing flesh, bodies in pieces or in sickeningly unnatural positions. We are empathetic and imaginative creatures; images and thoughts of such abnormal bodies make us flinch.

Sometimes the pain stirs us angrily into action; but sometimes it leaves us shrinking away in distress, taking refuge in blander descriptions, or pushing away the problem altogether.

The horror we feel feeds into our moral judgement of the perpetrators, but does not always determine it. Social relationships will also play a role (are they our boys or enemies—and how patriotic are we?), as will information we glean about their motives, if we can. The intensity of our disgust at perpetrators we dislike will be heightened by evidence of their power, physical aggression, and callousness, and particularly by cues which signal that their actions were unnecessary. Callousness may be terrible, criminal, wicked, and morally despicable, but our condemnation tends to be less vehement. Perhaps we recognize that callousness preserved our ancestors, and that we and our allies have been or could be callous in certain situations.

Sadism is different. A man who destroys an obviously helpless child, or rapes a woman prior to killing her—and especially one who laughs and jokes and boasts about it afterwards—brings down our moral obloquy. He may not think of his actions as sadistic, but we do. The battle for justification, waged by so many perpetrators and their allies, revolves around motive and necessity, around providing explanations which make their cruelty seem unavoidable—an unfortunate means, a side-effect, or a mistake rather than an enjoyable experience pursued for its own inherent pleasure.

The Nazis are today thought to typify sadistic cruelty. From their leaders' point of view, however, their actions were not gratuitous—except for the few deplorable 'bad apples' who enjoyed torturing and killing helpless people. In the sense defined above, they did not see themselves as sadistic. Rather, the extermination of the Jews and other enemies was a harsh necessity, a matter of survival. We see the Nazis as abominably cruel, but we must label the majority of their harm-doing, atrocious though it was, as callous, not sadistic. Their cruelty served other ends. Delight in pain was unmanly, unheroic, and unworthy of those selected to build the thousand-year Reich. It disgusted many of those who came across it—even within the murderous *Einsatzgruppen*.[35]

Callous and sadistic cruelty need differ not at all in their physical impact on their victims. The harm done can be just as appalling.

Nevertheless, we judge sadism, not callousness, to be the moral nadir of human atrocity. To see why, we must examine our two approaches to making such judgements—the pragmatic and the emotionally moral. What are they for? How do we come to have them? Those questions form the topic of Chapter 3.

Chapter 3

Why does cruelty exist?

If our all-too-brief look at the roots and development of warfare has taught us anything, it is how timeless and transglobal are the confrontations forced on us by the deeply embedded instincts that we aggressive humans have acquired through natural selection.

(Barry Cunliffe, 'The Roots of Warfare')

For nothing is evil in the beginning.

(J. R. R. Tolkien, *The Lord of the Rings*)

Discussions of human harm-doing often begin by contrasting the views of two great philosophers. Thomas Hobbes (1588–1679), with his famous description of life in a 'state of nature' as 'solitary, poore, nasty, brutish and short', is cited in support of the claim that humans are naturally violent and constrained by social structures. Jean-Jacques Rousseau (1712–78) is referenced for his concept of 'the noble savage', whose idyllic existence was smothered by the ruinous effects of civilization.[1] Thus the contrast between 'nature' and 'civilization' defines an opposition between those who see humans as naturally good but corrupted by society, and those who see people as wicked but reined in by group restrictions, at least most of the time.

The sharp division between what is social, or 'civilized', and what is natural may seem obvious at first. Look more closely, however, and problems arise. For one thing, 'natural' does not equate to 'good', despite the efforts of retailers to convince us otherwise. Malaria and earthquakes are natural; mosquito nets and earthquake early-warning systems are not. Humans developed civilization for good reasons, however much some of us may disapprove of the result.

For another thing, 'social' does not equate to 'human'.

Animal cruelty

> You'd figure the hawk for an isolate thing
> commanding the empyrean,
> taking his ease in the thermals and wind
> until that retinal flick, the plunge and shriek—
> cruelly perfect at what he is.
>
> (August Kleinzahler, *Anniversary*)

Many creatures lead complicated social lives, including displays of aggression in certain circumstances such as territory defence. Some—cats, mink, and polecats, for example—have long had the reputation of inflicting cruelty on prey. Chimpanzees, lions, and wolves seem to show cruelty to their own kind, ganging up on lone rivals to administer an agonizing death at low risk to themselves. The victims, in other words, pose no threat to their killers, while the latter show every sign of deliberation prior to the attack and excitement while it is taking place. Is this not cruelty?

Certainly we find it easy to make the judgement. Just as we see agency everywhere, so we instinctively use the same criteria to judge animal cruelty as we do for the human variety: deliberate, voluntary behaviour, causing foreseeable suffering to an innocent and undeserving victim. A notorious case is the footage, from the BBC's television series *Blue Planet*, in which a killer whale is seen catching and then apparently playing with a dying sea-lion pup by throwing it repeatedly into the air.[2] Whether we react with mesmerized horror or bravado, with mocking comments or hyper-rational analyses, we are drawn to see the orca's behaviour as cruel—despite our awareness that orcas are sea-lions' natural predators.

In doing so, we fall into the same traps which snare us in human cases, biased by our feelings for victim and perpetrator and hamstrung by our inability to grasp the perpetrator's motives. How much more innocent a sea-lion pup appears than a rat pup (yet both species are prey). How much more likeable dolphins are than killer whales (yet both species kill). How easy it is to think, because we can't see why the animal is acting as it does, that its motive must be a delight in torturing—the very motive we otherize as 'bestial' or 'brutal'. An orca playing with a sea-lion may be honing its reflexes, practising old moves or trying out new ones, or teaching the next generation essential life skills. Insofar as the sea-lion is capable of suffering, it undoubtedly suffers. What we do not know is whether the orca even understands the concept of suffering, let alone intends or savours it.

Killer whales are fairly distant relatives of human beings. What of our closer cousins the chimpanzees, whose deliberate and extremely violent killing reminds us of ourselves? If we think of cruelty in human terms, as unjustified voluntary behaviour which causes foreseeable suffering to the undeserving, then these attacks look very much like human cruelty, which is why they have attracted so much attention.[3] The more we learn about chimpanzees, the more we suspect that their capacity for high-level concepts, such as understanding the intentions of others, may be greater than previously thought.[4] They clearly feel both pain and distress, and seem to understand that other creatures also experience these feelings, implying some degree of empathy and the understanding of suffering. It may prove to be the case that chimpanzees, like humans, can take delight in beating a stranger to death.

Cruelty involves observable behaviour, and that behaviour can be described in non-moral terms. The amount of violence involved can even be quantified to some extent. One could ask of a chimpanzee attack: what was done? How much suffering was caused to how many for how long? Were the actions coerced? Could the goal have been achieved in a less harmful way? Cruelty, however, also implies moral questions: were the victims undeserving? Did the perpetrators' goals justify the means employed? Were they morally responsible for the act? Crucially underpinning these questions is the claim that it is right to subject the perpetrators to moral judgements. That is, they are assumed to be not just 'moral patients' (like young children, who are granted certain moral rights but are not held morally responsible for their actions), but moral agents, responsible for their behaviour and capable of sufficient understanding to know that they are doing wrong.

The question of chimpanzee competence in moral cognition is one for philosophers and primatologists. At present, it is probably fair to say that most experts would not attribute such high-level moral understanding to their primate cousins. When we call chimpanzee aggression 'cruelty' and react with horror, we are experiencing a basic moral reaction which derives from our assessments of our species and is then extended to others through anthropomorphism. If people behaved like that we would think them cruel. In other words, we *do* readily extend our moral judgements to other species, whether or not we are *justified* in doing so. More to the point, we condemn our fellow humans just as readily—and sometimes with equally fragile justifications.

Morality and free will

There is another even more fundamental issue here. Lurking like a black hole at the heart of every discussion of human atrocity is the question of free will, agency, and responsibility. Viewed in non-moral terms, our actions, however free they may feel, are in fact caused and thus in some sense inevitable. Moral responsibility has no place in this cold, grey universe, since it requires holding people to account who could not have done anything other than what they did (thus says determinism). The second, moral approach maintains a much stronger view of agency, seeing the active self as a real and independent entity rather than a construct or illusion. Moral free will holds people responsible for their actions despite their abusive backgrounds, subjection to propaganda, bigoted peer groups, and so on.

These two views of free will are extreme theoretical positions. In practice, as noted earlier, people sometimes accept determinist arguments and sometimes insist on moral agency. The battle is over which influencing factors should 'count' as causes—thereby discounting our moral responsibility for what we do—and which should not. Severely psychotic people, for instance, tend not to be held accountable for their actions these days; nor do people afflicted with brain tumours. At the other end of the causal scale, many factors, such as star sign, are usually taken to be causally irrelevant (perpetrators who claimed to have murdered because they were born a hot-tempered Aries would probably have more chance of leniency on psychological than astrological grounds). In between lies a legal and moral quagmire, best decided on a case-by-case basis.[5]

To glimpse the depths of this problem, consider some examples of influencing factors. One is obedience to authority, brought to prominence by the work of the psychologist Stanley Milgram.[6] The Nuremberg war crimes trials ruled out higher authority's ability to offer a moral discount—a literal 'get out of jail free' card—by dismissing the Nazis' defence of 'obeying orders'. Yet obedience still carries huge justificatory weight for many people (not just perpetrators), whether in the specific form of obeying authority or the more general form of obedience to one's imposed or self-adopted social role; that is, ultimately, submission to others' expectations. This was the kind of obedience demonstrated so alarmingly in the Stanford Prison Experiment.[7]

Another influencing factor is prior abuse. Whether you think severe childhood abuse should mitigate in cases of serial killing, for example, will depend on your culture, political position, and personal experience. Research suggests that most serial killers suffered appalling abuse as youngsters.[8] Does this mean that serial killers should be treated, rather than punished (or killed, in some societies)? As we learn more about our brains' sensitivities to mistreatment, especially during early development, the list of potential influences is likely to increase. What of psychopathy and the personality disorders, ideological indoctrination, social exclusion, brain damage, substance abuse? What of irresponsible parents and childhood deprivation? As we learn more about the causal weight these factors carry we will adjust their moral discounts accordingly. If everyone who took cocaine or was diagnosed with psychopathy inevitably turned out to be a murderer the case for a discount would be more obvious, but life is not so simple, and for many potential causes the appropriate moral discount remains uncertain. We do not yet know enough to set the tariffs.

Discounts and explanations

One of the problems with trying to work out what the moral discount for a given bad behaviour should be is that such calculations occur in many different circumstances. Legal operations to dole out society's punishments are only the formal tip of a very large iceberg. The media, friends and colleagues, and we ourselves pass moral judgements easily and automatically. Sometimes no actual perpetrator need exist; often our judgements have no impact on real perpetrators; frequently they are ill thought through, swayed by emotions, or presented with the aim of making the speaker look better to his or her associates. No wonder justice is usually portrayed as blind.

A second problem with moral discounting is that potential influencing factors do not all seem to use the same currency. Some of our explanations centre on agency: what philosopher Daniel Dennett calls 'the intentional stance', which sees behaviour in terms of motives and action-goals, beliefs, and desires.[9] 'The orca attacked the sea-lion because the orca needed to eat.' The human tendency to extend agent-based explanations to humans and much else besides forms a symbolic world alongside the physical, through which those equipped to build theories of mind can navigate. Our capacity to mentalize has given us the realms of gods and spirits, of hidden forces wishing us well or ill, of animals, plants, even mountains and tiny organisms—all with their own interpretable minds.

Scientific explanations of behaviour look very different. In evolutionary theory, the 'agent' on whom the explanations centre is either a species or a gene, not an individual. Evolutionary explanations refer to the 'deep' reasons provided by natural selection: 'the orca attacked the sea-lion because it has evolved to eat sea-lions.' (That is, orcas who killed and ate sea-lions in the past were better nourished, and left more descendants, than orcas who didn't, and the difference provided a selection pressure which, over generations, made orcas increasingly likely to eat sea-lions.) Current biological explanations give us accounts of behaviour which draw heavily on brain mechanisms—'he's violent because of dysfunction in his prefrontal cortex'—in which agency appears, if at all, as an illusion ripe for analytical dismemberment.[10]

Neurobiological theories of some kinds of human behaviour, such as the gag reflex which causes us to vomit, can get by without reference to agency (as we shall see, because a section on the gag reflex is coming up, so to speak, in Chapter 5). Theories of more complex phenomena like schizophrenia, or cruelty, lie at the interface between neuroscience and psychology and need to take some concept of agency into account, because agency is central to human function. We learn to split the world into agents and non-agents much as we learn grammar and prejudice— intuitively and with surprising ease. No explanation of cruelty which ignores agency, or sees it merely as an effect, will be complete.[11]

This also applies to explanations in the social sciences and humanities, which see behaviour in terms of an individual's social roles and location in hierarchies, groups, and institutions. Thomas Hobbes's interpersonal conflicts, Jean-Jacques Rousseau's smothering civilization, Michel Foucault's power systems, Karl Marx's historical and economic forces, and the group pressures explored by social psychologists are all of this type. Yet these causes, whatever they may be, do not seem to be the same kinds of things as genes or human agents. One is left wondering what common currency could possibly include such disparate factors.

As I have said, the view of cruelty presented in this book is necessarily incomplete. Anyone seeking authoritative instructions about which factors should have which moral discounts will not find them here; such detail would be premature. Instead, in the next section I will sketch out a 'working model' of a scientific approach to cruelty. Before we reach that stage, however, we already know at least three of its characteristics.

One is multiplicity. The model will require several kinds of explanations of what causes cruelty. So complex a phenomenon, encompassing everything from serial killers to kids throwing kittens on fires, is unlikely to boil down to a single formula, an '$E = mc^2$' for evil. Instead, multiple factors will mesh in certain times and places to produce the icy pragmatism of callousness, in other situations to evoke the fierce eagerness of sadism. Those factors will include social and cultural forces: peer pressures and the demand for obedience; ideologies, myths, and stereotypes; sudden economic change; and many more. They will include biological forces: drugs and hormones, inbuilt threat-response programmes, and so on. They will also include the evolutionary forces—what James Waller calls our 'ancestral shadow'—which shape our needs to compete and otherize.[12] All of these will interact with each other and with the individual's own unique history.

Another characteristic of the model concerns the question of how to define cruelty. I will use the formula developed earlier: unjustified voluntary behaviour which causes foreseeable suffering to an undeserving victim or victims. This can be expanded to include the further distinction between callousness (in which the victim's suffering is intended as a means to an end) and sadistic cruelty (in which the victim's suffering is the aim of the exercise). Any theory of cruelty, however, must address its moral aspects as well as its physical and social causes, because morality is an inescapable part of the concept's definition. That means, first, teasing out the moral logic involved, and secondly, explaining why human beings might have such morals in the first place.

The third feature which any model must possess—here are my colours, firmly nailed in place—is that it must be grounded in our knowledge of the human brain. Why? Because only in the brain do we find the common currency we need: the neural codes into which everything else can be translated. We already know that genes and chemicals affect how brain cells behave, although we are far from understanding the intricacies which lurk behind that deceptively simple statement. We also know that brain activity changes in response to far more intangible stimulation: beliefs, memories, desires, intentions, ideas. Thinking about one's role in the group, one's economic situation, political preferences, or sacred cause—all these involve changes in brain function. We usually see large-scale social forces like 'the economy' as operating at the level of entire societies; but (to state the obvious) those societies, like Soylent Green, are made out of people. A misogynistic stereotype, fear of recession, or belief in violent jihad exerts its dire effects because people's brains are changed by exposure to it. That is the underlying mechanism, and any large-scale theory which ignores it ignores a useful source of information, because brains can only behave in certain ways. Step outside of those limits, for instance by assuming that people act on the basis of logical decision-making, and your theory will eventually come to grief.

No book can fully reflect the complexity hinted at here. We must simplify, resorting to a working model of cruelty which can capture principles and incorporate data from multiple disciplines without (at least initially) getting bogged down in details. That model should explain existing facets of cruelty and make testable predictions. It should apply both to cruelty's worst extremes and to less serious forms of human harm-doing, referring back to the working definition of cruelty and setting cruel behaviour within a wider moral context. It will, however, be necessarily incomplete and quite possibly wrong. All scientific models are either one or the other, unless they are both. With these caveats, let us proceed.

A model of cruelty

> I am more convinced than ever that the capacity to perpetrate genocide is not limited to one culture or one people but is an inherent potential of the human condition.
>
> (Robert Melson, *Revolution and Genocide*)

HUMANS HAVE EVOLVED AUTOMATIC RESPONSES TO THREATS

Life on earth evolved in a world full of dangers. Predators and scavengers, poisonous plants and animals, horrible diseases, floods and lightning and earthquakes and volcanoes, even hostile members of one's own species: all potentially lethal. Human beings have felt threatened for as long as they have felt at all. Many threats, like guns or unemployment, are of evolutionarily recent origin, but some, like sharks and bacterial infections, affected species long before we emerged. The need to detect and respond to these variegated horrors provided selection pressures favouring the survival of individuals who responded quickly and effectively.

As humans evolved they had time to evolve threat responses. Pathogens, poisons, and predators changed only slowly; the threats of disease and of being eaten were ever-present. This consistency allowed organisms to develop specific 'programmed' threat responses— physiological changes, feelings, and behaviours which were triggered automatically by certain stimuli. Humans have kept these evolved suites of fast reactions, just as they have kept the instincts of predation and scavenging (as any observer of our behaviour in supermarkets knows).[13] We jump at sudden noises we know to be harmless, feel uneasy in dark places, and react to certain animals with fear and horror—whether or not they could actually do us harm. Abrupt movements startle newborn infants as well as children and adults. Even for those of us who never encounter a dangerous predator and have modern equipment to protect us from pathogens, the instincts remain. Shared history, not the actual degree of danger, is what matters. In Britain, more people die from paracetamol than spiders, yet arachnophobia is much more common than fear of painkillers.

FEAR-TRIGGERING THREATS

The kinds of threats which have been around long enough for humans to have evolved responses to them can be classified into three types according to the primary negative emotions they produce.[14] Fear-triggering threats (hereafter fear-threats) come from sources so powerful that they threaten existence itself, sources which may be too powerful to resist. Reality dominates; its helpless victims must either flee or submit and hope to survive. Natural disasters, like volcanic eruptions and hurricanes, still fall into this category. Many other formerly overwhelming threats mean less to us now, thanks to technological protection. Stripped of modernity's carapace, however, even the most physically powerful human is likely to be helpless against large animals, smaller creatures in large numbers, hard-to-detect, fast-moving, and dangerous creatures like stinging insects, and other uncontrollable hazards of nature (wild seas, slippery or shifting ground, and so on).[15]

These natural dangers do not exhaust the repertoire of fearful stimuli. Our ancestors were also afraid—with good reason—of members of their own species.[16] Human beings can be unpredictable and lethal. Moreover, some are much more powerful than others, whether because of physical strength, skill with weapons, or because their high social status gives them allies. Group members who could assess the power of others accurately, and whose fear of the powerful made them behave submissively, were less likely to provoke aggression from their social superiors, which made them safer and group relations more harmonious. Accurate detection of fear-threats, and appropriate responses, were thus selected for by social hazards as well as by natural ones.

When confronted with a powerful predator, a rat's instinctive reaction is to freeze and crouch. Internally, 'stress' hormones such as adrenaline are released into the bloodstream; the brain's activity changes, raising alertness and focusing attention on the threat; and blood (i.e. energy) supplies are diverted from low-priority organs like the gut to high-priority organs like the heart and limb muscles, in order to get the body ready to run or fight. The very similar responses in humans generate the thumping heart and wide eyes of terror without the need for conscious decision-making. Later inventions from natural selection's workshop, like the learned ability to inhibit one's initial reactions, can override fear-threat responses to some extent, although not completely. Self-control is fragile, energy-intensive, and liable to fail, especially under stress.[17] The more potent the threat, the more likely it is that old-established reactions will prevail.

Fear leads to long-term as well as immediate changes. The behaviour which led to the fearful event may be inhibited. Environmental features associated with the event may become aversive, as in phobias and post-traumatic stress disorder. The physiological mechanisms which mediate these responses evolved to have a powerful impact on behaviour and memory, because fear-threats typically arrive with little warning and demand immediate attention. If the threat continues to be anticipated for long periods without the danger actually materializing, sustained anxiety and chronic stress can cause physical wear and tear, psychological passivity, and depression. Many modern fear-threats, such as terrorism or job insecurity, are of this type.

ANGER-PROVOKING THREATS

A second category is that of anger-provoking threats (hereafter anger-threats), which typically occur when agents misbehave (for example, threaten one's resources or act unpredictably).[18] Their sources are powerful but not overwhelming: they can, with effort, be resisted and sometimes controlled. Whether physical obstacles, persistent scavengers, uppity rivals, or disobedient offspring, anger-threats announce reality's defiance and generate the aggressive urge to resolve the problem, reasserting control through vigorous attempts to force the unexpected into line. Anger primarily targets other agents, as these can be influenced to change their infuriating ways, but the human readiness to extend the intentional stance to all and sundry means that we often treat objects as agents which aren't. I know perfectly well that shouting at a machine which isn't working contributes precisely nothing to making it work. Nevertheless, I have been known to swear at recalcitrant computers, and I am not alone.[19]

Anger-threat responses, like those for fear, tend to be stereotyped and similar across species, suggesting a long evolutionary history. Cats confronted by a rival widen their eyes, snarl, tense their muscles, and raise the hairs on their back to make themselves look bigger and more dangerous; sometimes the challenger will rethink and retreat. Men about to fight glare, show their teeth, clench their fists, and hunch their shoulders for the same reason; both species use similar hormonal mechanisms. A defeated cat rolls over to expose its belly as a signal of submission, while a man losing an argument may spread his hands—the human equivalent, leaving his body open to attack—to show that he concedes the point. Bodily signals of dominance and submission are learned early in life, and can often be understood by other species (thus are dogs trained).

Anger provides the motive force for defending against controllable threats from others. Given that humans are intensely social creatures who spend much of their time and energy interacting with other humans, most anger-threats are unsurprisingly social: quarrels over status, resources, or access to mates. They are risks to status and power, not directly to physical survival. Anger and fear interlock in the management of intragroup relations, motivating the development of stable status hierarchies which minimize aggression between group members.[20]

Anger, like fear, can do physical and psychological damage if it becomes chronic or excessive. Many of the physiological changes which characterize the fear response, such as the release of adrenaline, also occur during anger, and indeed they can sometimes be hard to tell apart. In fear, behaviour which led to the threat may be inhibited, however, while in anger it may be reinforced. If the attempt at regaining control is successful, anger can be a very rewarding feeling.[21] Failure, by contrast, may lead to the same threat evoking a more fearful response in future.

DISGUST-EVOKING THREATS

Not all of nature's hazards are as obvious as an earthquake, a snarling tiger, a charging rhinoceros, or a furious rival. Sometimes the damage may only become apparent well after its source has been detected. Yet the harm done may be out of all proportion to its source. Even the biggest poisonous spider is small compared with humans, easily crushed, and quite harmless—unless you let it get close and it chooses to bite you. Rotten meat, 'off' milk, or other degenerating matter can't even bite you; neither can corpses, vomit, faeces, or open wounds. The danger comes from close contact—touching, or worse still, mistaking foul food for fresh and ingesting it. Contact transmits either poison or virulent pathogens, ready to divide and conquer, causing death, long-term damage, or incapacitating illness.

An infected body reacts with a physiological transformation, activating its immune defences, raising core body temperature, triggering thirst and sweating to flush out pathogens and poisons, and vomiting to evict the invader from the gut. Spitting removes dangerous material from the mouth; a wrinkled nose and averted face reduce the risk of airborne pathogens entering the body; and cleaning behaviour, particularly around wounds or sores, protects the skin. These activities need not be deliberately undertaken, although they may be intentionally used as social signals. Rather, they comprise the evolved responses which humans use for dealing with these kinds of threats. The concomitant emotion is disgust.[22]

Anger- and fear-threats are reactions to dangerous situations. Disgust-evoking threats (hereafter disgust-threats) are reactions to potentially dangerous situations. Disgust is essentially preventative, a 'keep away' signal. If distance is maintained by avoiding the disgusting object, the threat it poses will not materialize. If avoidance is not a practical option, removing or if necessary destroying the object should resolve the problem, assuming the procedures are such as to minimize the risk of contamination. Procedures for the avoidance, expulsion, and elimination of disgust-threats have become immensely ritualized in every human culture, because every culture, however safe and socially harmonious, must constantly deal with contaminating matter.[23] Postmodern man may serenely dominate his universe to such an extent that he rarely feels fear or anger except vicariously, but he will still produce repellent organic waste. He is also likely, at some point, to fall ill, whereupon he may well become disgusting to other humans he encounters. Visibly sick people are often laden with the kinds of features we learn to associate with disgust reactions (dripping noses, open sores, and suchlike). Presumably those early humans who avoided their diseased colleagues 'like the plague' were less likely to die from the diseases. Unfortunately for those thus categorized, foreigners who look different and people with visible blemishes or deformities can provoke the same reactions.[24]

The quest for physical improvement, cosmetic surgery and all, has sound Darwinian logic behind it, because some physical abnormalities result from either current infection, previous infection (that is, they indicate susceptibility or a potential carrier), or genetic defects—all of which can deter potential mates. Many differences from the norm, however, have no such implications; but natural selection is not always the most subtle of judges. Human beings need not be slaves to natural selection, gene-puppets driven by ancient urges (one reason why the more grandiose generalizations of evolutionary psychology should be taken with a sizeable helping of salt). Overriding instincts, however, takes effort.

Disgusting objects may threaten physical survival indirectly through the pathogens or poisons they transmit. They may also threaten status by transmitting disgustingness from themselves to the person who touches them. (Would you shake the hand of a man who had just a minute ago picked up some faeces, even if he had washed his hands during the interval?)[25] Their main threat, however, is to physical integrity. They penetrate bodies and change them in unpleasant ways, weakening their structure and spreading decay from the inside out. Until relatively recently the mechanisms by which disgust-threats achieved their corrupting effects were unknown—invisible miasmas, particles, or fluids, curses or spirits or *something*—and even now the causes of illness can be fiendishly difficult to find. Sly and secretive, hard to detect and thus hard to defend against, pathogens and poisons can be recognized only by their visible effects on organic matter. Signs of decay and damage provide the cues. The causative agents themselves might as well be incorporeal spirits, unless you have a microscope handy. The only guide our distant ancestors had in avoiding their perils was that poisonous organisms and decaying matter are often vividly noticeable (e.g. brightly coloured or malodorous). That, and a growing ability to learn.

LEARNING RESPONSES TO THREATS

Not all threats are 'programmed', in the sense of triggering automatic and stereotyped responses (which may or may not be inhibited by the individual). You may start at a sudden noise or flinch if someone sneezes all over you, but what do you do the first time you meet a new object? Here is where the bond between threat responses and their associated negative emotions comes in useful. If you eat the new object and then feel sick, you will thereafter regard it with considerable disgust—and disgust is linked to a suite of physical responses which have evolved precisely to keep you away from things which make you ill. By categorizing the new object as disgusting, you incorporate it into the set of objects which trigger disgust-behaviour: avoidance, expulsion (removing the object), or elimination (destroying the object), depending on which is easier, together with cleaning rituals for objects whose repulsiveness is only superficial. You have reduced a range of potential behaviours to a familiar, efficient set of responses, which will make life much simpler next time you encounter the object. As with all threats, the aim of the response is to restore control over your environment by negating the threat.

Humans learn by trial and error. If eating cheese makes them throw up, that's it for Welsh rarebit; if gooseberries once give them food poisoning they'll probably be left feeling sick at the thought of a gooseberry, never mind the thing itself. But most of what disgusts us we never go near, let alone consume. Instead, we learn from others by imitating, observing and discussing our feelings. As members of a community we have access to a wealth of accumulated information about disgust-threats, stored as cultural norms (for example, surrounding food) which we dutifully pass on to our nearest and dearest. An angry man and an angry cat are in many ways similar, but human disgust is uniquely sophisticated. We owe that sophistication to our affinity for group living and our language-laden capacity for culture.

HUMANS HAVE EVOLVED TO LIVE IN SOCIAL GROUPS

Human beings have evolved as social animals, finding safety in kinship groups.[26] Within a group, individuals competed for social status, which brought greater access to resources (including mates) and thus a better chance of raising offspring. Until the development of weapons a lone individual, however physically powerful, could not withstand either attacks by large predators or a concerted assault by other humans, so those individuals able to form alliances were better equipped to survive fraught confrontations than their less socially adept peers. Aggression against group members risked alienating potential allies, whose help might be essential for the aggressor's survival next time a predator came calling. Behaviour which damaged one member of an alliance risked bringing down the wrath of other members, and this distribution of power enforced codes of conduct which form the basis of all the moral systems we know today.

Cooperation, unlike intragroup violence, was highly beneficial. Obeying moral rules—the social 'covenants' proposed by Thomas Hobbes—incurred the costs of caring for other group members, aiding the weak and protecting the vulnerable; but carers could expect that, should they be injured or fall sick, they too would be cared for (reciprocal altruism). Successful communities used altruistic punishment (so called because the enforcer incurs the costs of punishing and gains no direct reward) against not only the overly aggressive but also the 'free-riders' who took without doing their share; even today, free-riders are a major source of workplace discontent.[27] Human groups, left to their own devices, seem to reach a natural maximum at around 100 to 150 members—larger, and some members won't know others—which, it is thought, reflects the importance of social reputation to our ancestors.[28] Cheating is much harder when everyone you are likely to interact with knows you're a cheat.

In a kin-based group reciprocal altruism is a win–win strategy, despite its costs.[29] Not only do you benefit directly, you also raise the survival chances for copies of your genes located in the gonads of your kin (especially if you can estimate the degree of kinship and target your favours accordingly). Even if some group members are not related to you, the direct benefits of reciprocity still hold. Long-term, it pays to be nice to the people you live with. At times, as noted in the Intro-duction, fashionable thinkers have derided morality for its 'bourgeois' conventions, its conservatism, its intolerance of difference, and suchlike; but from the Darwinian perspective moral codes appear to have been a stunningly effective 'good trick'.[30] Reciprocity and altruistic punishment can generate highly cohesive groups whose power, when their members act together, far outweighs the power of any human acting alone.[31]

The twin demands of dodging danger and dealing in diplomacy would have placed a heavy burden on cognitive resources, strongly selecting for those individuals canny enough to cope. At times, as evidence of genetic bottlenecks caused by population crashes suggests, survival must have been a close-run thing.[32] Yet some hardy creatures did squeeze through—and look at the results: today there are billions of us. (Evolu-tion's gloriously flourishing masters or the sapient fungus rapidly smoth-ering the planet: take your pick.) The survivors, our ancestors, achieved their Darwinian success not by piling on muscle to take on ever-larger predators, or by specializing in one ecological niche, but by developing the capacities needed for acting as part of a group. Paramount among these was the ability to guess what would happen next: both in the phys-ical world and in the increasingly complex and dominant social world. We conquered nature because we learned to predict it.

HUMANS HAVE EVOLVED SOPHISTICATED POWERS OF PREDICTION

The ability to use information about their bodies and environments to make predictions about future events and their likely consequences is one which numerous animals possess. Species survive by evolving two predictive capacities: expertise in judging how dangerous a threat is likely to be, and appropriate responses to different kinds of threat. An individual who cannot accurately predict the environment cannot learn to control it, whereas one whose predictions are accurate can adjust his or her behaviour to interact more effectively and safely with objects and organisms.[33] Predictions are essential precursors to control, and individuals who are better predictors are thus more likely, *ceteris paribus*, to survive and reproduce.

In a stable ecological niche the demand for predictive abilities, and the brain-power they require, can be low or non-existent. If the world around you is pretty much always the same, predicting it is an expensive waste of energy. Some species of polyp take this logic to its obvious extreme, retaining their rudimentary brain for only as long as it takes to find a rock to hang onto. Once securely attached, the resource-hungry organ is lost; it is no longer needed.[34] Humans, however, are generalists, exploring multiple niches, and they have taken the skill of prediction to startling heights.

Our brains make predictions continuously and automatically as part of controlling and monitoring our bodies. For example, you unconsciously predict where your arm will end up after you move it, which is a much more complicated computation than it may seem. You do this by generating short-term, constantly updated guesses—beliefs about where your arm will be—combining information about the way the world is with details of the size and direction of the movements you are about to initiate. These guesses are usually followed, almost immediately, by new information about where your arm has actually ended up, which either confirms, contradicts, or modifies your predictions.[35] Your brain compares the post-movement reality and the pre-movement belief about that reality, and adjusts the prediction to make it more accurate. The next time you reach for an object, your movement will be a little more fluid and skilful. When you are two months old, modifying predictions about arm position is essential because there is so much room for improvement.[36] Babies can focus intensely, and for far longer than adults or older children, on repeating a single activity: reaching for a toy or fingering an object. Adults, however, need make few, if any, changes to most of their guesses about how their bodies behave, and so are not normally aware of their motor predictions.

Some predictions, however, such as scientific hypotheses, concern complicated events occurring over much longer timescales. These require considerable effort to create and still more work to test. Some extremely important aspects of the environment (for example, its vulnerability to attack, flooding, lava flows) are of this demanding, energy-intensive type. Requiring lengthy study before predictions can even be made, let alone be checked against reality, they drain resources in the service of an uncertain outcome (as complex predictions may well not be accurate). The potential long-term benefits are huge, but given the short-term exertions, why would your averagely impulsive distant ancestor want to bother? Even today, the business of making the most specialized and effortful predictions, which we call science, is not an activity to which most human beings are naturally drawn.

CONFIRMING A PREDICTION IS INHERENTLY REWARDING

Here we see a conflict between the personal rationale, which requires an individual to conserve energy in pursuit of survival, and the evolutionary rationale, in which 'fitness' benefits (i.e. more offspring) accrue to those better able to guess where the floods or the lava flow will strike. One way in which natural selection has resolved other similar conflicts is by making the activity which generates the long-term benefits inherently rewarding. The obvious example is sex. The physical process of sexual reproduction has much in common with a disgust-threat and/or a physical attack: close proximity to others, body fluids, orifices, and so on.[37] Sex has costs for the individual: it consumes resources and leaves one vulnerable to other threats—including, if the sex is successful, offspring adept at draining parental resources. From this point of view sex is a repugnant waste of energy. To encourage reproduction, therefore, an incentive is required. Sex must feel good—so good that disgust, fear, outrage, and the dread of future impoverishment can be overcome. Arousal and orgasm supply the incentive, a currency of pleasure which makes sex inherently rewarding.

Human beings in general like to have sex, but their enthusiasm varies, both between and within individuals. On the whole men are keener than women, though both genders cover the gamut from revulsion through indifference to rapture. A minority becomes addicted to sexual pleasure, while sexual deprivation can be no problem at all or the biggest problem a person has ever faced, depending on the person. Like other intrinsic rewards, sexual satisfaction can be obtained in a staggering diversity of ways; shoes, dogs, dead bodies, and eye-watering pain are some of the more unlikely stimuli. Individuals restricted from fulfilling their sexual needs in one way may seek alternative routes to satisfaction. Moreover, the desire for sex conflicts with the desire to preserve physical and psychological integrity (including 'personal space'), creating a changeable balance of lust and disgust.

The evolutionary rationale suggests that prediction became inherently rewarding to human beings because those more motivated to make predictions were better players of the Darwinian game. Just making predictions, however, is not enough; indeed, it could be a dangerous waste of time. What matters is that predictions which come to be tested turn out to be accurate. The most effective rewards are those which come when predictions are confirmed, generating a sense of being in control which is as much a currency of pleasure as the warmth of a smile or the thrill of a flirtatious glance.[38] Incorrect predictions conflict with perceptions of the way the world actually is, and this conflict is aversive. That is, we are motivated (again to varying degrees) to avoid or minimize the feelings we get when dreams and reality clash.

An engineer designing a system to reward accurate predictions (that is, reinforce them when they occur, so that they become more likely to occur) would incorporate two mechanisms. The first is the one we have just considered: a signal—the sense of control—which tells the organism either that its expectations are in line with reality (a rewarding sensation) or that they have just been contradicted (an aversive feeling, triggering the need for greater control). The engineer, however, would also include a second mechanism designed to identify predictions which had either never been tested or not tested for some time, and prioritize the process of testing them whenever the organism had leisure to do so. Natural selection, however, is not an engineer. Boredom, curiosity, or science teachers may drive us to re-examine our assumptions from time to time, but in general the unexamined life is rather comfortable. In other words, we are built to prefer predictions which do not generate conflict, either because they are confirmed by their comparison with reality or because they are never tested. For the same reason, people who challenge our expectations may not always get a warm reception.

Features of the desire for sex—individual differences, gender differences, addiction in some cases, the aversive nature of deprivation, stimulus diversity, and the ingenuity prompted by restriction—are also found in the need to have control. Like sex, it has its 'balancing' desire: the urge to explore and enjoyment of novelty, whose aversive deprivation signal is boredom. Today's stable, predictable environments—physical and social—may fulfil the need for control to such an extent that boredom results, but human beings evolved in a world which was unpredictable, and sometimes lethally so. An adaptation which made prediction-making and prediction-confirmation inherently rewarding would have motivated our ancestors to seek out stability. That in turn would have reduced the amount of effort and time they needed to spend warily watching out for potential threats, providing them with additional resources of energy and time. Using this spare capacity to develop additional, more complicated predictions would have boosted their security still further (if you know the deadliest predators sleep all afternoon you can time your foraging accordingly, probably forage more efficiently, and certainly reduce the risk of being eaten). The result is a virtuous circle of increasing control over the physical environment. Moreover, the cognitive demands of making better predictions would have honed our ancestors' ability to manipulate information and to generate mental models of the world, conceptualizing events and objects which were not actually present at the time (e.g. past, future, spatially distant, or imaginary), and thereby gradually detaching thought from its dependence on sensation. From there, it is not so great a step to symbolic thinking, with all the opportunities for mastery that brings.[39]

CHALLENGES TO PREDICTIONS ARE INHERENTLY UNPLEASANT

In the previous section, I sketched the process by which predictions about the world, one's own activity, or that of other people are tested against reality and adjusted accordingly. That, at least, is the theory. In practice not all predictions are so readily abandoned. Tweaking a slightly inaccurate guess about one's arm position is one thing; abandoning a cherished scientific hypothesis is quite another. One might think that the need for control is fed by accurate prediction, and so ought to ensure that we keep our cognitive constructions properly subordinate to the diktats of physical reality. Not so, for two reasons. One is that our hunger for control does not demand that our predictions be actively confirmed, just that they remain unchallenged. It is the challenge, the signalled error between guess and truth, which drives the need for control. The second reason why mind and world diverge is that human beings, who are immersed in symbolic as well as physical environments, make many predictions concerning symbolic phenomena. Some of these, like the guess that the word 'fuckwit' is derogatory, can easily be tested. Some, however, cannot be settled one way or the other, such as the prediction that if hell were located Adolf Hitler would be found among its citizens.

As we shall see when we come to delve into the neuroscience, error signals warning of conflicts between predictions and reality are processed by regions of the brain, like the cingulate cortex, which also process pain.[40] Conflict feels stressful, like pain, and most people prefer to avoid it.[41] This makes evolutionary sense. If getting one's predictions wrong is so aversive, the unpleasantness can motivate efforts to be more accurate. The more important the prediction, the more intense the conflict when it is challenged.

When a person has invested considerable resources in generating a prediction and is heavily reliant on it being true, the very thought of the emotional and cognitive effort required to jettison it may be immensely painful and threatening. Rather than change their ideas to match the world, a person may seek an alternative way to remove the stressful feelings of conflict. Because the conflict signal is separate from the pre- diction, it can be adjusted without the cherished prediction having to be abandoned. This may involve coming up with new ideas to explain away the mismatch, or reinterpreting the prediction in some way, or reassessing one's perceptions of reality—not necessarily in that order. Scientists whose pet hypotheses generate predictions which are then proved wrong by experiments will typically first check the data, then check the set-up, then (if possible) run the experiments again. Meanwhile they will be frantically elaborating the original hypothesis, thinking up reasons why the experiments might not have worked. Only as a last resort will the hypothesis itself be abandoned.

Science's superlative effectiveness is due in no small part to the fact that, perhaps uniquely among humanity's competitive activities, it includes the idea of failure among its ideals. The need for control is so central that people go to extraordinary lengths to protect their beliefs, delusions, and other predictions about the way the world is. Scientists, like everyone else, aspire to be accurate, effective, and in control; but one of their beliefs about what they do is that hypotheses and theories are always revisable. Greater accuracy is possible; certain truth is not. Disproving a theory is commendable, as it means science is working effectively. Visiting speakers at scientific seminars are often introduced with a fulsome description of their achievements—grisly paeans, all too often—but the most sincere praise I have ever heard was for a senior scientist who had worked for many years on a theory, hugely influential in his field, which had since been proved wrong. He had not only publicly discarded his own theory, but was eagerly researching its replacement. An outstanding example to the rest of us? Yes, in science, though like all ideals it garners more praise than imitation.[42]

CHALLENGES TO SOME PREDICTIONS ARE INTERPRETED AS THREATS

Most of us, however, are not scientists, and even scientists can find it difficult to let go of precious ideas. Confronted with a challenge, we can often find an escape route, a reason why the mismatch is not really as bad as we first thought it was. Sometimes, however, the idea is so sacred and the confrontation so painful that the challenge strikes us as not just unpleasant but intolerable. Its very existence becomes a threat, requiring immediate attention and engaging the same threat responses we evolved to deal with far older hazards.

Human beings are not simply reactive. When the world is not as we desire or expect it to be we may adjust our predictions accordingly, or distract ourselves with other desires. But if that proves too difficult we may decide to change the world instead. If reality contradicts important beliefs, beliefs which it would be agony to abandon, then that threat of pain will be met by a threat response.

Just as natural selection occurs via genetic mutation, building on what is already present rather than starting afresh every time, so social evolution gave rise to new threats and reacted to them by co-opting older threat responses. The social, cultural, and symbolic challenges we face at present, like the objections to our way of life presented by those with very different ways of living, may at times become sufficiently threatening to evoke responses whose original triggers were quite different. Our threat responses to natural dangers were established long before the 'clash of civilizations'—long before civilization even began—but they were easily extended to symbolic hazards. Thus we have terrifying stock-market crashes, infuriating novels, disgusting images. Just flaccid concepts to most of us, most of the time, but to those directly involved the fear or fury or revulsion can be as intense and overwhelming as it would be in the face of more natural threats.

World-shaping in a crude sense happens all the time, of course; even bacteria change the rocks on which they grow. But prediction-based world-shaping, in the sense of changing physical reality specifically in order to match some abstract conception of how things should be, is a rather more complex capability. It has given humans unprecedented control of their environments, allowing them to make their lives increasingly safe, luxurious, and comprehensible. It has also, however, left them vulnerable to symbolic threats which cause no physical harm, but can nonetheless be perceived as highly dangerous. This discrepancy between symbolic threats and actual physical danger lies at the heart of human cruelty.

CRUELTY INVOLVES A THREAT RESPONSE

Thinking of cruelty as unjustified behaviour emphasizes the importance, and the difficulty, of matching threat responses to the threats which trigger them. Justified harm-doing can be accepted as defence or punishment. Problems arise, as we have seen, when perpetrators, victims, and third parties disagree about justification. Yet even when there is consensus in principle there may still be practical difficulties, because even though human beings have been dealing with social situations for thousands of years, there is still plenty of scope for getting things wrong.

As the environment's unpredictability diminished and human control over it increased, social unpredictability became a more significant factor in human existence, and social threats became increasingly important. (The nearest most of us pampered Westerners get to hunter-gathering these days is the local supermarket, and the biggest immediate threats to our survival are probably drivers in the supermarket's car park.) Other people, particularly strangers, have the annoying habit of not behaving as one would want or expect, aggravating the need for control and provoking aggression.[43] Interests may conflict, sometimes irreconcilably, such that others may threaten one's status, physical resources (such as territory), or even survival. Living with others will therefore create a selection pressure favouring individuals who can judge the social threat level accurately and respond efficiently. Aggression against potential future allies—or mates—must be finely calibrated. Too little, and you risk losing status; too much, and you risk group opprobrium. Either way you risk humiliation, and possibly damage, if you misjudge your opponent's power and determination.

The risk is very real, because the calibration mechanisms are far from perfect. Animals are not infallible when it comes to judging threats and how to respond to them—dogs achieve nothing by barking at passing fire-engines, though the noise does recede, so perhaps they think they do. Humans are not infallible either. It can be hard for even an experienced adult to know where to draw the line between warning someone off and terrorizing them. Computationally, assessing social threats is so hard a problem that despite all those years of practice we still get such judgements wrong at times. If your least favourite workmate makes a snide remark, the knowledge that it is only a remark may not prevent you having the kind of responses—raised heartbeat, muscle tension, clenched fists, and the urge to smash his face in—which helped your ancestors beat off physical attacks. Our threat responses are not always sensibly matched to their *agents provocateurs*—in others' opinions and sometimes even our own. And, unsurprisingly, mismatches tend to err on the side of excess. In our species' past, overestimating threats may often have been wasteful or damaging, but underestimating them was likely to be worse. Catastrophic misjudgements do happen, but most of us, most of the time, are skilled enough at theory of mind to smooth over tensions before they escalate. In a book about cruelty, it may be worth recalling that the vast majority of human interactions brings little suffering or none at all.

Abstract symbolic threats are different. An idea may take physical form in a painting or manuscript, but it may exist only as spoken words, breath on the air. Yet if that idea contradicts someone's strongest beliefs, they may see it as more of a threat than loss of status, physical damage, or even death. We know this because people sometimes choose to die, or suffer agonies, or both, rather than give up their beliefs. Emotions are rarely clear-cut, and a person confronted by the threat of a challenging idea may experience anger, fear, even grief and shock. But consider the type of threat such challenges pose. Ideas are dangerous to members of a community only if they risk becoming accepted by and incorporated into that community; as long as they stay in the dangerous outside world they can probably be ignored. In other words, the threat is not that the community will die (for which the appropriate response would surely be fear); nor is the community's status necessarily at risk (though it may be, in which case members will react with anger). The problem arises when the challenging ideas stand in such stark contrast to the community's core beliefs that accepting them would mean changing its very identity. Like sources of infection—a metaphor often applied to ideas—challenging notions seem incorporeal, spreading with horrifying speed to corrupt a group, or a person, from within. Given these characteristics, the obvious threat response is that of disgust.[44]

Disgust demands rejection, whether by avoidance, expulsion, or elimination of the contamination's source. People who voice objectionable ideas, like people who show signs of leprosy, evoke the instinctive demand for separation. If we can, we avoid them; provided, of course, that avoidance is not too inconvenient (for example, it may look like withdrawal to third parties, thereby making us lose status). Otherwise we expel the unwanted challengers from among us; or if that is not possible we may resort to physical destruction. As far as we are concerned, all this is self-protection. Others, seeing a leper, may agree that separation is essential and our insistence upon it entirely justified; but leprosy causes visible signs of illness which dangerous ideas do not confer. Unless they hold our threatened beliefs as dear as we do, third parties may fail to comprehend our disgust response. To them our actions will seem unjustified and cruel.

Summary and conclusions

Human beings do harm for many reasons: to save their lives, to preserve or enhance their social position, to teach the young, to enforce reciprocity, and so on. Morally speaking, these justifications are often sufficient to evade the charge of cruelty, especially when the threat is of a clear and present danger to life, limb, group equilibrium, or public morals. But some forms of harm are used to enforce social distance in response to threats which are neither clear nor present, and may not seem particularly serious. When those threats are symbolic, the need for separation may be unconvincing to third parties, especially if they do not share our beliefs. What to us is defence against a deadly threat to our identity—defined symbolically, not physically—to a distanced observer, or victim, is cruelty.

Later chapters will explore callousness and sadism, their origins and characteristics, in detail. To do that we first need to look at the mechanics of cruel behaviour. All cruelty shares one common feature: the medium through which it is expressed. Causal webs are woven in human brains. To get a deeper view of cruelty, therefore, means plunging our attention into a sea of neurons, the soggy, fatty mass from which cruelty is born. We need to ask about behaviour, since cruelty is something people do: 'Men begin with acts, not with thoughts.'[45] How do our brains take a superabundance of stimuli—perceptions, memories, beliefs, desires, emotions—and, quite often, turn it into a really rather smart response? That problem, the problem of action, is the topic of our next chapter.

Chapter 4

How do we come to act?

What source was there back then, save for our overelaborate nervous circuitry, for the evils we were seeing or hearing about simply everywhere?

My answer: There was no other source. This was a very innocent planet, except for those great big brains.

(Kurt Vonnegut, *Galápagos*)

Cruelty is, at base, about action. To understand what makes people cruel, therefore, we need to investigate the sources of cruelty (and every other human behaviour): our own nervous systems. They are the means by which information about our bodies, personal histories, and environments is transformed into the experiences we feel from moment to moment, the thoughts, beliefs, and memories we have, the emotions and moods and non-conscious stimuli which sway us, and the motor signals which trigger our every move. Other-ization and empathy, anger and pity, desire and disgust, the urge to take risks and the need to be in control: these conflicting urges meet and clash in the cerebral arena. The brain takes those urges and many others and weaves them into the mesh of emotions and ideas which every human being builds simply by living. Out of this neural froth emerge both great tenderness and extraordinary cruelty—sometimes displayed by the same

90

person within the space of a few moments. This chapter, the most technically difficult in the book, will examine what current understandings of human brains are able to tell us about the sensorimotor alchemy which can turn a vicious idea into a vicious action.

Sense and similarity

Brains are phenomenally complicated. A truism of course, the first thing we're normally told about our cerebra, followed by lashings of gruesome statistics involving very large numbers of cells, synapses, atoms in the known universe, or whatever. Grasping brain statistics, however, is a task brains themselves find challenging. What does all this complexity actually mean?

For a start, it means that no two events in the world—even very simple events, like flashes of light—will have exactly the same effects on you and your behaviour. To see why not, we need to start at the sensory beginning. Signals from 'out there' in the world (vision, hearing, touch, taste, smell) or from 'in here' in the body (visceral, hormonal, immune) come into contact with the nervous system in a variety of ways. Some are detected by receptors—specialized cells which can react to photons of light, the pressure waves of sounds, scent molecules, and so on by sending an electrical wave along a nerve fibre. So for the vision required to detect a flash of light, the retina at the back of each eye is stimulated when photons crash into it. The optic nerves are excited, and the electrical pulses they transmit spread like ripples through the brain, first to deep-buried areas like the superior colliculus ('little roof') and the lateral geniculate ('knee-shaped') nucleus of the thalamus, then to the visual, parietal, temporal, and frontal regions of the overlying cortex (see Figs. 2 and 3).

Yet that does not mean 'the same thing' happens for each flash, because each of those brain areas contains many millions of interacting neurons, only some of which will care about the first flash; and some of these will be distracted by other goings-on within the brain from responding to the second flash. Even if we take a brain signal out of context and pin it down to a single neuron irritated by a single stimulus, the response to a second, identical stimulus may be different.[1] At any given instant, neurons add up the number of signals reaching them and react

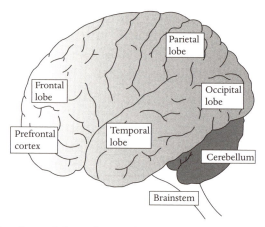

FIGURE 2. A lateral view of a human brain, with the four major divisions of cortex (the frontal, temporal, parietal and occipital lobes) labelled. In a lateral view, the brain is seen 'sideways on', with only one of the two cerebral hemispheres visible. Here the front end (anterior) is towards the left and the back end (posterior) towards the right. Thus the occipital lobes are posterior, the frontal lobes anterior. Regions can also be subdivided into, for example, anterior and posterior temporal cortex. Two major subcortical landmarks, the cerebellum and brainstem, are also labelled (landmark labelling is done so that figures can be compared).

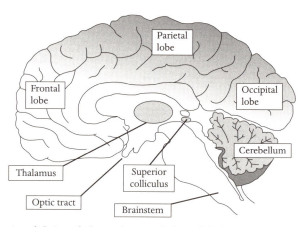

FIGURE 3. A medial view of a human brain, with three of the four major divisions of cortex (the frontal, parietal and occipital lobes) labelled. A medial view represents the brain as if it had been split down the middle, with one half-brain removed so that the inner surface of the remaining half is visible. The cortex forms a wrinkled layer over the subcortex. The superior colliculus, the thalamus and the optic tract—which contains nerve fibres from the eyes—are labelled. Two major subcortical landmarks, the cerebellum and brainstem, are also labelled.

with a signal of their own if the balance of their inputs exceeds a certain level, the cell's electrical threshold. Since every neuron connects to many others, inputs may well change between the first flash and the second, in response to unrelated events, like breathing. Neurons, furthermore, can change their behaviour on the basis of experience, which is why people can too. The brain which responds to a second event was changed by its response to the first event, so the two responses will not be quite the same. For our purposes, as neuroscientists or psychologists or simply people in a hurry, they are often similar enough—but that is because we choose not to look too closely.

This surface similarity, hiding deeper difference, applies to all our mental furnishings as well as to the cells which implement them. When I use the word 'cat', for instance, you know what I mean because of your past experience—direct, or indirect via images and stories. You know that cats are usually furry, that they purr and miaow, that they often chase and kill small creatures, that their young are called kittens, and so on. To this extent you and I share the same concept of 'cat'; but we don't need to dig much deeper to diverge, because our different histories have coloured that concept with details unique to us. My archetypal feline mingles a long-dead family pet with the current incumbent, throws in Buffy-the-calico-kitten-from-next-door, and adds a dash of innumerable literary and media cats, from Kipling's solitary prowler to Eliot's Macavity. Yours may be friendlier, saner, or altogether less positive; what it will not be is identical to mine.

And yet we can connect. We can talk about cats, we know their ways; they have been part of our culture for centuries. They are real; out there; in some sense independent entities (notoriously so, in my experience). They continue to exist, we presume, when we are not aware of them, and they have stimulated human sensory (input-receiving) brain regions in fairly consistent ways across many different times and places: by purring, scratching, yowling, and the like. The two ideas of what the word 'cat' means in two different brains—or the same idea in the same brain at different times—are thus as similar and different as two oak trees. From a distance they appear identical, but when you look closely you see the variation in branches and leaves.

Why does this matter for a study of cruelty? For several reasons. First, 'cruelty', unlike 'cat', is an abstract and complex moral concept many of whose examples are not subject to easy consensus. Two politicians may have diametrically opposed ideas about whether, say, the techniques used on prisoners in the 'war on terror' are cruel. Secondly, this very flexibility makes cruelty a problematic concept. People in the English-speaking world who wish to communicate with others learn to call a certain kind of small furry animal a 'cat'. They could make other noises on the relevant occasions, but the social costs of being considered an idiot so far outweigh the boost to personal autonomy that the decision stands: the English word is 'cat'. To teach you its meaning I simply point out a cat, or find a picture of one; you understand what pointing means, and how to read pictures, so because there is something there to indicate we can link up. But for cruelty I could point at anything from a mother smacking her toddler to a man being tortured. Worse, for each example I chose, there are or have been people who would maintain that I am quite mistaken (the smack is educational, the torture necessary punishment). Yet all of us use the concept of cruelty.

Concepts and networks

We readily imagine concepts as networks: a kind of mental knitting, with perceptions and prior knowledge stitched together by association. That metaphor accommodates the flexible, ill-defined nature of many of our ideas. Precision is a matter of degree: the more precise a notion, the more similar will be the stimuli which evoke it, the associations it evokes, and the moves we tend to make in response. Concepts also vary in their richness. My concept of an adenosine trisphosphate molecule is a scanty construct compared with that of a molecular biologist who has studied these crucial cellular components in detail, because although I need it for my work I do not use it every day. Cell biology being a minority pursuit, most people need no concept of ATP at all (though they wouldn't last long without the molecules themselves).

Ideas are usually pragmatic things, responsive to our needs, remarkably ready-to-hand (so to speak) when we want them, tucked out of mental sight the rest of the time. We form them and link them together very easily; indeed, humans devote extraordinary amounts of time and energy to playing with concepts. Often we do not examine them in detail; but we can hone them into rarefied tools stripped of extraneous associations—like the naked x and δx of calculus. Then we can use them as symbols with which to build castles in the air: the rococo castles of mathematics or theoretical physics, or real castles—and plenty more besides. Alternatively we can use words as poets do, like a harpist's fingers plucking the concept-networks, triggering rich waves of associations which will overlap but be unique for every reader or listener. Mostly we do something in between, varying the way we use our concepts depending on the situation, our mood, the role we're playing, and many other factors.

Achieving such flexibility and keeping track of how concepts relate to each other requires some mechanism capable of providing more information than any book and more dynamically powerful than any computer. Such a mechanism must offer rapid access, so that concepts can be retrieved when needed. It must also offer an easy way to bind concepts together, or to push them further apart as differences between them become apparent. The facility to change the links between concept components is a pre-requisite, if only to correct mistaken first impressions; concepts are learned, after all, and may need adjustment. The mechanism must tie concepts closely to the objects, events, and processes in the world which they represent, so that if some aspect of reality changes, the concept which represents that aspect changes too; connections to the world are therefore essential. Finally, it must allow us to zoom in or out on concepts as we please, using them superficially when the focus is on similarity or speed of use, and investigating the rich webs of associations when we want to emphasize our uniqueness, explore an idea in depth, or simply take our time.

To get an inkling of how superabundant and variable those associations are, you may like to try listing the associations which come to mind when you think of a word, for example, 'cruelty'. Two people's attempts are shown in Table 1, for two minutes-worth of free association. As you see, they overlap but are quite different.

Table 1. shows word lists generated by free association in response to the word 'cruelty'. Two participants, A and B, heard the word 'cruelty' and then listed every word or phrase which came to mind for up to two minutes. The first 30 unique words are listed (in practice, A and B found the task so difficult beyond this total that few additional words were generated). Six words, listed (in BOLD font) in the upper section of the table, were cited by both A and B. These words reflect cruelty's association with harm-doing, and especially with large-scale violence and extreme, gratuitously-inflicted suffering.

The lower section of the table shows words unique to A (left column) and B (central column). The words are grouped according to theme: situations associated with cruelty; perpetrators and their motives; the otherization of cruelty as evil and inhuman; situations and emotions associated with cruelty, particularly with respect to victims; and the opposition to and punishment of cruel behaviour.

WORDS COMMON TO BOTH PARTICIPANTS		
'aggression', 'genocide', 'sadism', 'suffering', 'torture', 'violence'		
PARTICIPANT A	PARTICIPANT B	THEME
Holocaust, wars	*concentration camps, cult, Middle Ages, Russia, war*	Situations where cruelty occurs (extreme)
	playground, schoolgirls, teasing	Situations where cruelty occurs (less extreme)
perpetrator	*Hitler, male, Slobodan Milosevic, Stalin*	Perpetrator
desire to hurt, indifference, ruthlessness	*malice*	Perpetrator motives
atrocity, demonic, dirt, evil, incomprehensible, inhuman, monstrosity, savage, terrible, wickedness, vile	*dark, fairytales, ogres, viciousness, wild*	Otherization of cruelty as distanced from us
despair, dismay, horror, rage, terror	*chains, dungeons, oppressive, powerlessness*	Feelings associated with cruelty and its victims
human rights	*The Hague*	Opposition to cruelty
punishable	*punishment*	Punishment

To the brain

The mechanism we use to manipulate concepts is the human nervous system (the brain and its neural tendrils), and it is ideally suited to the task.[2] It delivers the flexibility, the rapid access, and the skilful monitoring of reality so smoothly that we rarely notice—unless, of course, it malfunctions. For every one of its activities that we become aware of there are many more which carry on unrecognized. As you read this text parts of your brain are maintaining your posture, adjusting your heartbeat and breathing, or receiving and acting on information about your body's hormone levels, your immune function, and the state of your internal organs. Other areas are interpreting and reinterpreting the messages which reach you via your optic nerves, guessing which word comes next, allowing you to move your head without the text slipping out of focus. Still others are readying layer upon layer of meaning for each move your eyes make, tempting you to break the thread of the text by pausing to explore a tangential association—and much, much more.

TIME AND COMPLEXITY

What we do not have here is the behaviourist caricature of stimulus-followed-by-response: direct, predictable, robotic. As anyone who has tried to study human beings knows, their behaviour varies even on the easiest tasks. As the tasks become more complicated people take longer to respond, which gives more areas of the brain more time to chip in with their contribution to what that response should be.[3] This is why much low-level military training is about simplifying tasks down to their motor essentials: move, aim, fire, and do it fast. Ideally, targets must be obvious—preferably armed men in uniform, clearly attacking—so that the task of selecting each individual target can be performed as fast as possible. The weight of more difficult decisions— 'Are these genuine enemies? Is our response justifiable?'—is taken by people higher up the command chain, freeing soldiers on the ground to do their dangerous work without having to pause and consider every move.

Complicated stimuli slow our responses down considerably. There is so much more processing to do before deciding how to act. Compare the task of reacting to a simple visual image, like a white circle on a black background, with an image of, say, a busy street. Instead of a rapid neuronal consensus—'the interesting bit's on the left, nothing much happening elsewhere'—we have a huge number of committees in different brain areas, all talking at once: arguments over shapes, colours, and movements; guesses about object identity and location; reports on what we are expecting to see and whether or not an item needs to be prioritized. That we act at all is a tribute to how much more efficient brain committees are than human ones. One effect of all this interactivity is that the more complex the input, the greater the role of 'high-level' contributions to the output: expectations, prior knowledge, beliefs, and so on. This immense complexity ensures that the laws of human nature will be statistical: while humans quite often behave in similar ways, their precise reactions in real-life situations are inescapably varied. People may be described as simple-minded, but nobody is ever simple-brained.

Sometimes we genuinely cannot make sense of what we see, as in the impossible buildings depicted by M. C. Escher or ambiguous images like the Necker cube (Fig. 4). In such cases, we use our prior experience of buildings, or cubes, to generate an expectation of how they should look. In the case of the Necker cube, that gives us two conflicting cubes, one facing towards us, the other away; and that, in alternation, is what we actually see. As elegant experiments have shown, the flipping from one view to the other is mirrored by the activity of groups of neurons in

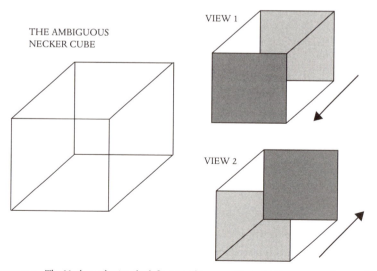

THE AMBIGUOUS
NECKER CUBE

VIEW 1

VIEW 2

FIGURE 4. The Necker cube (on the left), an ambiguous object which at any one time can be seen in either, but not both, of two conflicting views (which are shown on the right).

visual areas of cortex.[4] As our perception changes, so does the behaviour of our brain cells.

We are, of course, all individuals. However, there is more room for the differences to emerge—in a sense, we are *more* individual—when we are dealing with complicated stimuli and have the leisure to respond as and when we please. That gives us time to listen to the opinions of more of our neural committees—opinions we tend to ignore when we are in a hurry—which may contain relevant information about previous experience. Someone who fills in a questionnaire about immigration, say, at their own pace may take the time to recall particular arguments they have heard, immigrants they have met, or other reasons why such and such an answer might be preferable. Someone answering under time pressure is likely to produce a much more stereotyped response, an effect exploited by demagogues for years. A key ingredient in much cruelty, therefore, is the imposition of severe psychological pressure on perpetrators. Whether through the physical stress of military training, anxiety due to economic or political crisis, or even simple time pressure, the sense of needing to act, and the sooner the better, makes all of us readier to otherize—and otherization makes us more likely to be cruel.[5]

The overflowing brain

Why does not having time make such a difference? The answer lies in how brains work. You may be familiar with the view of neurons as information processors, switching on or off in response to electrical signals. You may also know the metaphors of telegraphs, telephone exchanges, computers, and the Internet which this perspective has dragged in its wake. Over the years these notions have done an excellent job of misleading excitable young researchers into thinking that any day now they'll figure out this ball of sludge and build a better one. Yet analogy is not identity, brains are not computers, and no amount of talk about 'information processing' is going to change that. We are nowhere near to figuring out the brain's complexities. So where does that leave us?

One way of improving the analogy would be to build computers which changed their outputs depending on whether you dunked them in tea or vermouth. Neurons are constantly bathed by cerebrospinal fluid (the brain's bathwater). Substances in the blood, like alcohol, can seep into this fluid, changing its molecular makeup, and thus affect the brain. In short, every neuron triggered by a nerve signal reacts depending on its chemical context, within and beyond the cell itself. That context varies with the body's state of being as well as with the activity of other brain cells. Thus, even neurons processing the most abstract of cognitions can be disrupted by a sudden change in body chemistry, whether that change comes from food, drink, or drugs, infection or illness, or a brain-commanded surge in adrenaline. We will return to this key point in Chapter 5, when we consider emotions. For now, we must simplify in order to unravel how actions are born.

THE INWARD RUSH

When signals flow through peripheral nerves to reach the central nervous system (that is, the brain and spinal cord), they are routed through a series of subcortical brain areas to the cortex. There, like streams pouring into a river, they add their input to the neural activity responsible for generating behaviour. Along those input routes, whenever one nerve cell, say in the sciatic nerve, connects with another in the spinal cord, the signals passing along the route are changed in a way which depends on what else is happening around them. Put simply, stronger signals (those sent more frequently by the sensory receptors) tend to get stronger while weaker ones tend to fade away, becoming lost in the background noise.

As each signal pours into the brain, along with all its fellows, the level of neural chatter rises. Any individual neuron can synapse (make a connection) with thousands of others, so an incoming signal which only involves a few neurons may find itself a small fish in a gigantic pond. Moreover, at each synapse it may or may not manage to activate the cells on the other side (again, this depends on what other messages those cells are getting). If it does not, it has effectively failed to reproduce. Even if the signal does get transmitted through the synapses, its chances of affecting behaviour are low unless it happens to stand out from the crowd.

That distinctiveness may occur for a number of reasons. First, the signal may simply be very persistent, like a committee member who insists on repeating his point again and again until someone takes notice; it may also involve a great many neurons. The fact that military training, cult indoctrination, many forms of torture, and even everyday learning involve lengthy repetition is not coincidence. Neurons respect persistent sensory input, changing their behaviour in response to importunate signals to let them pass more easily next time.

Secondly, some signals are especially distinctive, or 'salient', for humans, because it paid our ancestors to take particular notice of them. Signs of body dysfunction, possibly caused by something which could threaten the observer, are an example. Once, travelling by bus, I saw people gathered round a fallen cyclist at the kerbside. She was bleeding, and I was astonished by how blazingly red the blood was. I don't remember much else about the incident; the salient stimulus grabbed my full attention—even though the situation posed no actual threat.

Thirdly, the signal may be altogether novel, like a new member who, just by being new, is granted a maiden speech in front of the committee.[6] A novice killer, seeing his first corpse sprawling at his feet, may find himself unable to look away. We are biased to take more heed of new events; even tiny babies look longer at new objects than familiar ones. Known features of the environment are often predictable, and predictable features are both safer and less interesting. Humans who feel safe seek out novelty, to differing degrees, and find it intrinsically rewarding.

Finally, the signal may be unusual, like a controversial planning application which stirs up intense debate. Unusual signals involve familiar components, with a twist: those components don't fit together as expected. Stressful demands to justify one's claims, statements contradicting old beliefs, new perspectives on old problems—all of these refer to ideas already acquired, but insist on doing something peculiar with them.

GATED COMMUNITIES

These properties of signals—persistence, salience, novelty, lack of 'fit' within the status quo—stand out from the crowd because they conflict with what the person is expecting. As noted earlier, the brain continually generates expectations—predictions of what sensory inputs should come along next—based on previous experience of which events tend to go together and knowledge of what has just happened. Sometimes those expectations are very strong, as when you hear a well-known tune and find yourself humming along; you could carry on the tune if its sound were to stop. Expectations are like filters placed across the flow of incoming information. Signals which match up pass through quickly and unremarkably; signals which don't match are more likely to be weeded out at an early stage. Psychologists call this confirmation bias: people's greater readiness to accept ideas which fit with what they already believe.[7]

The stronger the expectations, the stronger a non-matching signal has to be to get through the sensory gates. You may be barely conscious of humming along to a tune, but if the music wrong-foots your expectations, deliberately or in error, your attention will abruptly focus on your auditory input—if the clashing signal is strong enough to be detected. Unless you are a trained musician, you are unlikely to notice a sour note from a viola in the thick of a symphonic *grand finale*; but you have to have a real tin ear not to wince when a singer mangles a familiar tune, one reason why karaoke can be such hell. Visual illusions and many magic tricks work on the same principle: set up a strong expectation, weaken the impact of the inputs which signal that the expectation is incorrect (for instance, by distracting the person), and lo! the misled brain registers what its own predictions said it would, not what has actually happened.

Most of the signals which flow into our brains never make it to consciousness.[8] Like river waters filtering through a succession of dams, much of the input at any given moment may be blocked at early stages in the neural pathways. Some, however, spills over to the next stage, where it merges with all the other events currently affecting that particular region of the brain, and some of that neural impetus spills over again . . . and so on, from sensory through motor processing until thought spills over into action like water pouring into an overflow pipe. The more complicated the stimulus, the longer this process takes.

Human brains are not passive entities. Their function is not merely to link up inputs (stimuli) to outputs (responses) in the most direct way possible. Instead, they generate behaviour which is not entirely stimulus-driven. That is, although it still reacts to changes in the world, it also takes account of other factors which need not be 'present to our senses' as current stimuli. Memories, beliefs, emotions, and moods provide an internal environment, unique to each individual, signals from which shape the response to signals from the world (and vice versa).

Your mental environment has been developing since before your birth. It reflects the gigantic, hugely improbable, utterly inimitable conjunction of events which made you who you are—*that* meeting or accident on *that particular* day; the book you read as a child which changed your mind about something important forever; people who hurt you or were kind to you, who may have also hurt or been kind to others, but not in exactly that way at exactly that time. This universe of difference in your skull is not an archive, its contents hidden unless you deliberately retrieve them. It influences your day-to-day behaviour, making your actions and choices yours alone.

Any object you come across causes some, but not all, of the cells in your nervous system to become active (start firing off signals to other cells). As the signals reach the brain there is division of labour, with different areas of the brain, and especially the cortex, specializing in different types of signals. Light on your retinae activates cells in the occipital lobes, towards the back of your brain; sounds stir up the temporal lobes at the sides of the brain, and so on (see Fig. 2). Within these overarching regions there are many further subdivisions (visual cortex appears to contain over thirty different areas, tuned to prioritize edges, movements, colours, faces, and so on), and within the subdivisions are hordes of individual neurons.[9] For any given object, the result is a shifting pattern of neural activity which is unique to that object on that particular occasion. As noted earlier, no two brain events are exactly the same. Similar objects, however, or the same object appearing at different times, will activate many of the same neurons, causing similar patterns to spill across the brain.

This similarity on different occasions can itself be thought of as a signal. Patterns which last for a while, or frequently repeat, involve gatherings of neurons whose firing, over time, is highly correlated (that is to say, the neurons are likely to fire together). Amidst the random noise of everything else going on, correlated activity patterns stand out. As we saw earlier with incoming information, the nervous system has mechanisms to strengthen distinctive signals and weaken the surrounding noise of less prominent activity patterns. (The biblical statement that 'unto every one that hath shall be given' could aptly be appropriated for neurons.[10])

How does the sharpening of the signals and suppression of noise occur? Through synaptic plasticity: adjustments of the junctions at which neurons interact.[11] For example, if neuron B tends to fire when neuron A is active, the synapse between them 'strengthens' every time a nerve impulse passes from A to B. This means that B becomes more receptive to signals from A, more likely to fire when A fires. Hence the activity of A and B becomes more highly correlated (they become more likely to fire together), and the patterns of which they are part become more distinctive. Greater synaptic strength gives a pattern a louder voice on the neural committees whose votes will produce the next action.

THE 'FILING CABINET'

One can speak—and brain researchers often do; it's a useful metaphorical shorthand—of synapses 'storing' information. ('Coding for', 'modelling', and 'representing' are similar linguistic devices.) What this does not imply is anything like a library filing system, photograph, computer, scribbled aide-memoire, or box of records.[12] All of these, unless they have been damaged, offer perfect retrieval; the text, image, or programme which has been stored remains the same as it was when it entered the system. It also requires interpretation by a user: filing cabinets do not understand their contents. However, applying *that* notion to the brain leads rapidly into nasty philosophical tangles ('if you need an interpreter in your head, doesn't your interpreter need another interpreter in its head?'), and so is best avoided.[13]

Instead, what it means to say 'the brain stores information about (or represents) the world' is that an input signal, caused when some feature of the world is detected by the nervous system, leads to a correlated pattern of neural activity which changes the synapses between the neurons involved such that the next time that input, or one very like it, occurs, the probability of evoking a similar pattern will be higher than it would have been otherwise.

This is why people use the shorthand—and talk about brains as if they were filing cabinets.

Let's elucidate. To be able to see a cat, we rely on the following:

- light bouncing off Buffy-the-calico-kitten hits our retinal photoreceptors;
- nerves at the back of our eyes are activated by our excited photoreceptors;
- after a complicated journey, this produces a extremely intricate and constantly changing pattern in our brains;
- this pattern has much in common with other patterns which were evoked during previous encounters with Buffy;
- it has somewhat less (but still quite a lot) in common with the patterns evoked during previous encounters with other cats;
- it has even less (but still some) overlap with patterns evoked by dogs, tigers, pictures of cats, the word 'cat', and so on;
- it doesn't have much in common with patterns caused by biros, gas bills, etc.;
- and so on, for any concept you care to choose, although for abstract concepts like cruelty even the strongest patterns are likely to have less in common with each other than is the case for cat-related patterns.

In other words, if we encounter an object, a unique pattern of neural firing will occur. The more we examine the object, the longer the synapses involved have to increase their strengths, and the more highly correlated the pattern becomes. The next time we encounter that object, the same pattern (roughly) will be likely to occur again. Even if only some of the neurons are initially activated (for instance, if part of the object is hidden or missing), the likelihood is that they will quickly all become active. This is because the stronger synapses which connect neurons participating in the pattern allow signals to flow through them more easily than synapses which were not changed during the previous encounter. Signals flow most readily along channels set up by prior experience, which is why we are slower to respond to a novel object than to a similar one which we have met before.

Libraries by themselves can store masses of information, but they can't categorize their contents (that's what librarians are for). The brain can and does. If, for the first time ever, a non-Muslim meets someone who describes herself as 'Muslim', the patterns of activity in the non-Muslim's brain—in response to the sound of the word and the features of the speaker—will fire together as described above, strengthening the synapses between them so that the sounds of the word 'Muslim' become associated with this particular Muslim's features. Some of these latter will be common to every human being, so their patterns will be active whenever the non-Muslim meets any person, or looks in the mirror. More distinctive are the new or unusual features of appearance or behaviour— for example, a niqab or hijab, accent, skin colour, or language—and it is these which will become most strongly associated with the word 'Muslim'.

In future, human beings who share one or more of these features may be provisionally categorized, on first acquaintance, as Muslims, whether or not they identify themselves as such. Each individual produces a unique pattern in the observer's brain, but the patterns overlap and have many synapses in common. That overlap is the basis of the category labelled 'Muslim': common elements of the patterns tend to strengthen more than elements which differ from person to person. The more Muslims he or she encounters, the better the non-Muslim will become at distinguishing them from other people—and each other—that is, the richer and more nuanced the non-Muslim's 'Muslim' category is likely to be. As a method of Muslim-detection, it should also become more accurate as the person learns more facts about Muslims, for example, that belief has more than skin colour or accent to do with making someone a follower of Islam.

In theory, the very nature of a task determines the form of its fulfillment. Where there is the will, there is also the way, and if the will is only strong enough, the way will be found. But what if there is no time to experiment? What if the task must be solved quickly and efficiently? A rat in a maze that has only one path to the goal learns to choose that path after many trials. Bureaucrats, too, are sometimes caught in a maze, but they cannot afford a trial run. There may be no time for hesitations and stoppages. This is why past performance is so important; this is why past experience is so essential.

(Raul Hilberg, *The Destruction of the European Jews*)

There is one important caveat to the idea that our categories tend to become more accurate. Insecurity, stress, or a sense of threat can interfere to favour simpler views of the world. This brings us back to an earlier theme of this chapter: the idea that a brain under pressure to act quickly may simply not have time for all the relevant neurons to be activated before the decision to act has taken shape. To see why this might be the case, we start with a soundbite: in brain terms, more leisure means more noise. A brain in a hurry needs decisions from its various committees *now*, which requires at least some consensus within the committees themselves. The more divided their voices, the more those outputs look like random noise, and the less likely they are to affect the activity in other brain regions. Hasty brains will therefore react on the basis of the strongest neural patterns available. Since these are often caused by current events, this tends to produce more stimulus-driven behaviour. A leisured brain, by contrast, can take account of much more information before the command to act finally spills over into behaviour. This is why the choices we make under pressure can often differ greatly from the choices made in less stressful situations.[14] It is why people whose usual lives involve very little cruelty can, under certain conditions, act with a cruelty they would not have thought possible before, and cannot understand when confronted with it later.

What at the neural level makes this happen? As noted earlier, signals entering the nervous system pass through synapses, which can be thought of as pathways routing the signals from sensory input to motor output. At each stage the incoming signals are compared. For example, experience trains the brain to develop inhibitory synapses between neuronal networks carrying incompatible signals—those which do not normally occur together. (An inhibitory synapse is one which *lowers* the chances of a signal passing through it; if neuron A fires neuron B will be *less likely* to fire.) This keeps the conflicting neural patterns separate and distinguishable.[15]

In sensory terms, incompatibility reflects the familiar properties of objects. Real buildings cannot look like Escher's structures; we know this because we've seen thousands of real buildings. Up and down, wet and dry, black and white, cruel and gentle—the world is full of oppositions, and we perceive it in this binary fashion because of activation and inhibition: the neural choice to fire or keep quiet. In motor terms, incompatible signals are those which prompt incompatible actions, such as moving one's left hand forward and backward simultaneously.[16]

The presence of inhibitory synapses means that, for instance, the signalling of a neuron 'interested in' (activated by) flashes of light towards the upper left of a visual scene will tend to squash the activity of other neurons whose interests lie elsewhere.[17] Neurons which prefer flashes towards the lower right will be most affected; less incompatible signals (e.g. from the mid-left region) will be less affected. How much squashing occurs depends on how intensely the neuron responsible signals. Mathematical modelling shows that when inhibition is present one or a few 'strong' neurons can rapidly silence all the competition.[18] The more evenly distributed activity levels produced by more complicated stimuli take longer to sort out a winner. They that have, gain; they that haven't, lose.

Inhibition occurs at every stage along the sensorimotor route, reducing noise (weaker signals) and sharpening stronger signals. As signals reach the brain, however, many new sources of noise pour into the fray, and the field of potential competitors expands enormously. To cope with this, there is evidence that neurons in one part of the brain can inhibit neurons with different interests (while encouraging the activity of their

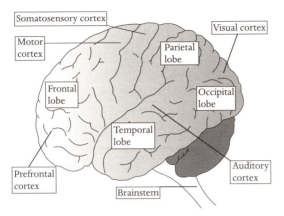

FIGURE 5. A lateral view of a human brain, with the four major divisions of cortex (the frontal, temporal, parietal and occipital lobes) labelled. The primary cortical areas which process vision, hearing, body-sensation (somatosensory cortex) and motor control are approximately indicated. The prefrontal cortex is also shown, as are two major subcortical landmarks, the cerebellum and brainstem.

more 'like-minded' colleagues) not only in their own localities but in more distant regions as well. This cross-linking further helps select the dominant signals from their weaker fellows (a process the neuroscientist Gerald Edelman has labelled 'neural Darwinism').[19] When it comes to areas such as the motor cortex, which generate the signals controlling movements, the same interactive rules apply, resulting in paralysis until the movement-related neurons can sort out among themselves which of them have won. A comment on the efficiency of this process is that we relatively rarely freeze in indecision.[20] Actions are normally as fast and fluent as they are for two reasons. First, the underlying synapses have been strengthened by the repetition of practice. Secondly, brains divvy-up their decision-making. By the time signals reach motor areas of cortex much of the work of reconciling their conflicting voices and competing urges has already been done.

The joy of flow

Thinking is like sport: it can be fun but it does require commitment. Making the effort to get some regular exercise is hard work, especially to begin with, and conscious, reflective thought is similarly demanding of time and energy. The much more common default, for healthy adults at least, is smoothly expert perception and behaviour, carried out with minimal conscious intervention. Acting seems to 'cancel out' the neural patterns which triggered the action, preventing them from interfering with future movements; and this helps stop patterns from becoming strong enough to reach consciousness.[21] Brains are naturally lazy, adept at learning easier ways to do things: effort uses precious energy in an already resource-hungry organ.[22] Thinking can interfere with this acquired expertise. Focusing on what your feet are doing as you walk is a good way to make yourself fall over them; focusing on the details of plot or acting is a good way to ruin a movie (especially a blockbuster). In both cases, the pleasure comes from not thinking, just 'going with the flow', whether that flow derives from your rhythmic stride or the sweep and swirl of special effects across your cortex. People speak of 'losing themselves' in the moment, of being swept away by the flow of events—especially when those events are highly salient and the pressure to act intense. Members of a group which has become a mob, for instance, may literally not think consciously about what they are doing at the time, making post-hoc justifications much more difficult. Some human actions are consciously planned, their implications thought through and their risks assessed, but most involve no such deliberation.

The consciousness we experience during 'flow' does not seem to be the same as full self-consciousness, but something much more basic and perceptual; some philosophers call it 'awareness'.[23] Conversely, we are often most conscious of ourselves *as selves* when we are suddenly brought up short by some hitch or difficulty. In that instant between recognizing the problem and beginning to solve it, the brain's expectations clash with the news from reality, and the mesmeric smoothness of 'flow' judders to a suddenly self-conscious halt. While flow is generally reported to be enjoyable, its interruption, and the feeling of mental conflict which can ensue, can be experienced as jarringly unpleasant. One of the areas of the brain involved in responding is the anterior cingulate, which as I noted in the previous chapter is also involved in pain perception (although we do not yet know whether the same individual neurons are involved in registering both pains and conflicts).[24]

When a clash occurs, the effect at the neural level is a blockage, with 'top-down' inputs (signalling the brain's predictions) contradicting 'bottom-up' inputs (signalling the current news of the world). As we have seen, under such conditions processing takes longer to generate an output. In the rest of the brain, however, signals still flow, reaching areas of cortex, notably in the frontal and prefrontal regions, which in turn send signals back to earlier stages in the processing pathways, boosting active neurons and damping down less active ones.[25] This speeds up the process of selecting winners and thereby helps to ease the blockage. It is these 'feedback' connections which allow higher-level predictions about what should happen to interact with sensory inputs. Depending on the other signals reaching the frontal lobe at the time (e.g. signals 'coding for' the urge for a quick response to that pushy salesman's offer), some frontal and especially prefrontal neurons will become more active, and their filtering of activity in other brain regions will thus increase, reducing competition and speeding up the passage from input to output. The price is that signals from some networks activated en route will play less of a role in determining the overall response. Most are irrelevant, so this is no great loss. Some, however, may carry useful messages that do not have time to assemble themselves and get noticed—like memories of what happened the last time you fell for an offer that seemed too good to be true.

Summary and conclusions

The message of this chapter can be summed up as follows:

> *Brains are really complex.*
>
> *No, I mean* really *complex. Far more complicated than we imagine. I can't possibly convey how intricately difficult they are. Just trying to think about it does my head in.*

On a more serious note, we have begun to make sense of the neural maelstrom. For one thing, we have learned that our superabundant brains are more than a match for the constructs we need them to instantiate. Our ideas, beliefs, perceptions, and cherished symbols—which can seem as impossible to pin down as a handful of mist, or as solid as granite—must be expressed through a medium of equal flexibility. That medium is, at base, statistical: the denizens of the mind are *not things* but *patterns of neural activity* which spring into being in response to a given input.[26] The strength of an idea conveys the readiness and frequency with which it comes to mind; its importance in and dominance of a person's life, and the extent of the bodily changes which take place when that particular pattern occurs (changes which we can interpret as emotions).

For familiar inputs the synaptic connections are strong, and the probabilities involved can be as near to certain as makes no practical difference. The face of a loved one will reliably conjure storms of excitement in your fusiform gyrus, amygdala, and other regions (unless of course your fate involves a neurological disorder like Capgras syndrome, prosopagnosia, or the dreaded Alzheimer's).[27] For less familiar or more complex inputs, the probabilities of retrieving the exact pattern triggered by the original stimuli are likely to be lower. More chance of error, less chance of confident assurance, less stability over time.

The strongest patterns are the ones that seem most like things to those who have them, because they are as stable and recurrent as the patterns prompted by objects we can see and touch. These are the certainties which can feel more real than the real world, more central than the self, more important than existence. Weaker patterns seem fluid and insubstantial. Every pattern, however, surges out of the same neurochemical swirl. Mental furnishings they may be, but the apt analogy is that of lighting and atmosphere rather than curtains and chairs.

This view of what neurons do is one of the hardest and most unnerving ideas in modern science—as well as one of the weirdest and most beautiful—so at the risk of boring you I will reiterate it yet again. The metaphor of brains as computational storage devices, soggy PCs, is misleading. Instead of mental 'hardcopy', what we have is a causal connection which results—on the whole—in similar neural patterns when we perceive similar events. These neural patterns in turn tend to generate similar behavioural responses. We are not talking certainties here, although much of what goes on in the brain has been made highly probable by three useful features of our particular universe:

1. It is physically stable over time. Most objects in the world, from trees to tables, stay put, most of the time, do not abruptly change their shape, size or colour, do not liquefy or disintegrate or turn into Christmas crackers.
2. Our brains and bodies, in general, appear to be pretty similar. Korean eyes, Fijian livers, Zulu median nerves, and Latvian hippocampi seem to work the same way as their American, Dutch, Ghanaian, or Thai equivalents.
3. Many of the situations we encounter, and much of what we do in response to them, involve huge amounts of repetition, strengthening the causal links which underlie familiar perceptions and behaviours. How many times this week have you picked up an object, brushed your teeth, breathed?

We have also learned that deciding to act, whatever it may feel like, is not a unitary event conducted by a single Überneuron, or even a Central Executive Committee of neurons. Contrary to popular myth, everything which makes us human is not necessarily done by the prefrontal cortex, and that includes making choices. As you move your gaze across a picture, multiple areas of your brain are repeatedly guessing what it is you are looking at, investigating how you feel about it, choosing the place in your visual scene most worthy of closer attention, estimating how the eye movement to that point will affect the signals coming from your retinae, conjuring associated memories, and also performing a host of other tasks irrelevant to your appreciation of the picture—except insofar as they helpfully keep you upright, focused, and alive. Signals are filtered and background noise smoothed away as activity pours through the brain, from region to region and back again, so that by the time the intention to move has grown strong enough to tug at the sleeve of consciousness the movement itself has probably already been selected.[28]

Time is of the utmost importance when generating behaviour. A leisured person motivated to explore alternatives and a stressed person wanting only to get finished will have very different patterns of brain activity. The pressured individual will make more use of the strongest networks available, paying less attention to weaker, conflicting ones. He or she will be less likely to override initial impulses, more likely to disregard information about consequences or moral prohibitions, less likely to resist the suggestions or commands of others, and more likely to show stereotyped behaviour. Thus, a man who feels threatened may react aggressively even when he has good reasons to inhibit the violent response.[29] Either the inhibitions are not strong enough to smother the pro-action signals, or the suppression comes too late: by the time the inhibitory patterns have been activated the response has already begun. It is no coincidence that severe stress, whether emotional or physical, inevitably precedes the worst extremes of otherization and cruelty.

Repetitive learning also speeds up actions by strengthening the underlying neural patterns, making them faster and harder to inhibit. Like the programmed responses involved in reacting to threats, hunting, or having sex, they acquire such influence over behaviour that they can evoke responses 'automatically'. Short, swift paths through the brain become widely used, while longer trails are less commonly travelled, as if neural signals were lazy hikers adept at choosing the route of least resistance.

How easy the path is to travel, moreover, depends on who else is travelling it as well as on the route itself. A stimulus which activates an already active pattern can get a faster reaction than a stimulus which must stir up dormant neurons or, worse, inhibit their active competitors. Research shows that activating the relevant neural patterns before a stimulus is presented, by priming people with imperceptible cues, can give such patterns more influence over behaviour and make reactions faster.[30] With respect to, for instance, moral inhibitions, this means that unless the inhibitory patterns are active while the stimulus is being processed, they will have no voice in the neural committees, and thus will fail to influence the decision to act.

There are two implications worth noting here. First, a man may have all the moral education society can give him, show a clear understanding of his culture's ethical principles, act with charity and kindness to those around him, and still become a torturer or killer. Indeed, he may learn to kill babies without ever losing his morals, although if he is thoughtful he may find it difficult to readjust to normal existence afterwards. Moral qualms are useless unless they are used, active participants in the decision to act.

Secondly, activating part of a neural pattern increases the likelihood of activating all of it. This matters, because imagining or talking about doing something, watching someone do it, and doing it oneself all involve overlapping patterns which use the same brain regions.[31] Consequently, even mild otherization primes people for aggression, whether or not it explicitly encourages them to act aggressively. To think about doing something cruel is to take a step along the otherizing path which leads to cruel behaviour. Whether the next step is taken will depend on how the person reacts to the thought of being cruel. He may accept the thought unchallenged as his, as 'self', a fragment of the constant swirl of mental activity which feels either pleasant or no more unpleasant than other everyday mind-business. If so, there will be little interneuronal discussion, and the thought's underlying neural pattern will tend to strengthen. When it is next activated, the person will be that little bit more likely to cross the threshold into verbal expression— perhaps at first to close friends, then more openly, until he is smearing it across the Internet and being praised by like-minded others for doing so.

If, on the other hand, the thought triggers unpleasant feelings and/or the co-activation of patterns associated with 'non-self', the resultant neuronal conflict will prompt inhibitory signals from other brain regions to block the flow (the cerebral equivalent of sticking out a foot to trip someone up) until the conflict is resolved. The process of conflict resolution will either label the thought as acceptable to 'self', rendering it a legitimate contributor to the person's behaviour, or reject it as illegitimate, 'non-self'.

If an otherizing thought is expressed, or an otherizing action carried out, which incurs swift social retribution, it will become associated with anxiety and the stressful sensation of mental conflict, leading to inhibition of future actions (because the thought of doing them is unpleasant). Every step of otherization, encountered for the first time, involves an often-tentative testing of social boundaries: will this action bring punishment or reward? Each step also builds on those which have already been rewarded, such that each new reward or punishment is compared against an accumulating 'weight' of feel-good factors—the rewards which the person has learned to expect. Consequently, additional rewards and punishments become less effective as otherization proceeds. Like bad habits, viciousness is most easily halted early on.

Because related actions are represented by overlapping patterns in the brain, repeated activation of even mild otherization makes more extreme behaviour easier to trigger. This is why perpetrators who have been exposed to otherization for many years without necessarily committing violence, like the 'ordinary men' described by Christopher Browning, can adapt so easily and quickly to the task of killing.[32] It is also why people used to violent cultures, like gang members or the Khmer Rouge cadres, kill for, in our view, tiny and trivial reasons. Otherization and the social acceptance of violence have lowered the threshold required to trigger murderous aggression.[33]

We tend to think of voluntary actions as being accompanied by free and obvious choice on the part of the actor, as if he or she sat down quietly in a corner and listed the pros and cons before deciding whether to go ahead. This emphasis on rational choice is particularly strong when we think about perpetrators of cruelty. The reasons for not acting, we say to ourselves, were so compelling that reasonable people surely would have desisted. Blood, screams, empathy, moral revulsion: what could be more powerful?

The answer, unfortunately for victims, is that all sorts of motives, from fear to greed to the simple compulsion of having begun to act, can temporarily conquer a person's inhibitions. The choices which lead to acts of cruelty, though free, may not always be obvious, to the actor or others. That is to say, the relevant motivating signals may not be sufficiently distinctive, in neural terms, to be remembered at all, let alone extensively justified in advance. When that person is under severe pressure, as perpetrators often are, with the immediate demands of the situation providing motives to act which dominate the neural conversation, the person may not have either time or inclination to pay attention to conflicting urges—especially if hesitating or expressing qualms is liable to bring down ridicule or group wrath. In this climate, situational constraints (fear of going against the group consensus, for example) can exert much more effect than they would normally, and motives which in everyday life would remain no more than momentary 'bad thoughts' can come to prominence in a way which astounds both perpetrator and observers.

The fears of those who scream 'reductionism' at neuroscience can only be based on failure to grasp the intricacies of brain function. Reducing humanity to puppets on stimulus-driven strings would certainly be morally reprehensible, but that is not what neuroscientists do—however much governments might wish they did. The more we learn about brains, the more we are staggered by their complexity, forced to confront their variability, delighted by their beauty in form and function, and awed by their capacity to ground human uniqueness. But *that is all they are*—the raw material which allows a baby to transform itself into Gandhi, or Stalin, or you. To speak a message you need a voice; to think one, you need a living brain; but that does not mean the message 'is just' the sound waves, or the neurons firing. It is a thank you, a curse, or a goodbye. Which of these it is depends on you and your circumstances— that is, on factors both within and beyond the brain.

In the next chapters we must therefore reach beyond the skull to consider the webs of meaning into which all human creatures are woven. Those webs are social and symbolic, and their power to bind us derives from the fact that they are part of us. What we say and do, the symbols we cherish, and the roles we play define our identities as human selves. Just as our bodies define us physically as separate organisms—or sometimes unite us through empathy, imitation, a comforting cuddle, or sex—so the beliefs which populate our cognitive landscapes, and the way we feel about them, define us as like or unlike, pushing us together or apart. We are motivated toward self-defence in response to threats, whether the self in question is virtual or physical. But whereas a physical threat may be obvious to all, a symbolic threat may be threatening only to us and those we care about.

To unravel the causes of cruelty, we need to explore both beliefs and emotions. Beliefs structure the relationships involved and channel a perpetrator's desires into actions. Emotions fuel the urge to act. They are the force behind every act of cruelty, so let us begin with them.

Chapter 5

How do we come to feel?

Frequently we hear the passions declaimed against by unthinking orators who forget that these passions supply the spark that sets alight the lantern of philosophy

(Sade, *Juliette*)

The story so far

In Chapter 4 we examined how brains create behaviour, picturing brain function in terms of the ebb and flow of neural activity in pursuit of smooth, efficient actions. This allows us to see choices not as occurring at one decisive point (where the soul steps in to make the Überneuron fire), but as emerging from the interaction of many 'neural committees' in the brain. Conflicting neural signals are resolved through filtering and inhibition, such that stronger signals tend to win through, suppressing weaker ones at every stage of brain processing. Learning, stress, and pre-existing expectations (e.g. due to priming the person in advance) can make responses easier or harder to evoke.

At any one moment during the process of deciding to act (or not), multiple patterns of neural activity are busy resolving their differences. Some are associated with facilitatory (pro-action) signals, some with inhibitory (anti-action) signals, and some are irrelevant, neutral or background noise. Roughly speaking, brains weigh up the relative strength of facilitatory and inhibitory signals in a neural vote which determines the output. To understand a decision to act cruelly, therefore, we need to consider both the interplay of pros and cons, signals promoting and inhibiting action, and the timing of their passage through the brain. That timing depends on their relative influence: the strength of the underlying neural patterns. Many factors can change a pattern's strength, but in this chapter we will focus on one of the most important: emotion.

Sources of strength

We will return to the concept of 'strength' when we come to look at beliefs (in Chapter 6). For now, suffice it to say that the strength of any pattern of neural activity reflects its salience, its importance to the person within whose head it ripples. This in turn reflects the persistence and distinctiveness of the input signals which generate the pattern, and hence what else is going on in the brain at the time. Salience also reflects a pattern's rich or meagre associations with other patterns: like people, some patterns are well connected and others socially marginal. In addition, salience depends on the type of association: that is to say, on where the neural activity comes from. There are three types of sources: the world, the brain, and the body.

THE WORLD

Peripheral sense organs respond to the presence of objects external to the body. At the back of our eyes we catch light; our cochlear hair cells flex in the currents of sound waves; our skin is full of tiny cellular machines standing ready to sound the alert when we burn or are prodded, stroke velvet or scrape against concrete. Smell and taste, with their subtle chemoreceptors and analyses of texture, complete the picture of the traditional five senses. These provide external signals to neurons. Patterns of brain activity which rely primarily on external signals track reality closely, changing quickly when it changes, so that our models of the world are kept up-to-date.

THE BRAIN

The second source of neuronal activation is input from other neurons. A cell in your primary visual cortex, for instance, may receive signals from neurons in the lateral geniculate nucleus of the thalamus. These in turn take their incoming, 'feedforward' signals from the retinal surfaces in your eyes, via the optic nerve (see Figs. 3 and 6). Numerically, signals from other neurons make up a huge proportion of the whole. Even as early on in the system as the lateral geniculate, estimates suggest that for every synapse transmitting information from the eyes up to the brain there are ten 'feedback' connections from cortical regions sending information from the brain to the thalamus.[1] What we see is controlled, modified, interpreted long before we are aware of seeing it. As we

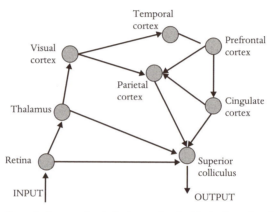

FIGURE 6. A schematic diagram of major brain areas involved in the formation of a movement (the example used is the fast, flicking eye movements known as saccades). Saccades are generated via a series of overlapping input-output pathways, each longer and more complex than the last (shown here in very simplified form). Areas involved at an early stage (like visual cortex) are heavily influenced by sensory input. Areas involved in more complex processing (like the cingulate) are more influenced by information from other cortical regions.

The fastest saccades are generated when light stimulates the retinal surfaces in the eyes to send signals to the thalamus, which connects to the superior colliculus, and directly to the superior colliculus. This can trigger an eye movement if the signals are strong enough (e.g. if the stimulus is very simple, or the person is expecting the stimulus and knows in advance where to look).

Slower, but still automatic saccades occur when signals from the retina do not trigger the superior colliculus immediately. The visual information, which is simultaneously relayed via the thalamus to cortex, then has time to reach visual and parietal cortical areas. Parietal cortex also sends signals to the superior colliculus, either reinforcing (perhaps sufficiently to trigger a saccade) or modifying that area's 'opinion' as to where the next eye movement will be.

Meanwhile parietal areas are also signalling to other areas in the frontal and temporal lobes, including areas of temporal cortex involved in recognising objects as well as prefrontal and cingulate cortex. These in turn add their 'votes' to the neural consultation in the superior colliculus. This allows time for more leisured decision-making. Saccades which take this long to happen feel more voluntary; they can be blocked and directed.

proceed along the processing arc from stimulus to reaction, through visual, parietal, and frontal cortex, we find the influence of sensory input shrinking still further. Neurons in 'higher' areas, such as the prefrontal cortex and cingulate gyrus, are on a longer and looser sensory leash, their firing patterns heavily influenced by signals from other cortical areas.

Patterns which rely on signals from other neurons need not have much to do with the way the physical world is. They may instead track regularities in the social world of other people's behaviour, such as word meanings or status hierarchies. These are not obvious, in the baldly physical sense in which a fire or a stone is obvious, but fires and stones, independently extant though they may be, become real to us only insofar as they evoke activity patterns in our heads. The more stable and predictable our environment, the stronger the patterns triggered by its features, and the more real we take those source's patterns to be. (Until, of course, we grow up and encounter philosophy and science; and even then, the doubts those two mischievous teachers whisper in our ears rarely engender more than superficial scepticism.) Whether the regularities they track are solid and palpable or manifest only in thoughts, words, and actions, we evolved to take strong patterns as evidence of stable, predictable realities.

And so they often are—but not always. The presumption that strength reflects reality can be misleading, because strength can come from other sources than sensitive monitoring of a placid universe. Strong emotions may have nothing to do with the way the world is, as we shall see; yet they can make a belief look like eternal truth, even when it is danger-ous nonsense. To evolution, of course, ontological concerns are of no relevance except insofar as they affect fitness. If a side-effect of emotions' power to spur protective action was that certain ideas were mistaken for realities, what did it matter as long as the genes were still passed on? And for most of humanity's deep history it didn't matter, because emotions too were closely bound to the situations which evoked them—until we learned to love and protect ideas.

But that is for Chapter 6. First, we must examine the third source of neurons' stimulation: that wellspring of feelings, the body.

THE BODY

We tend to think of our senses as extracting information from 'out there', but news from 'in here' is equally important, and our body is full of sensors providing just that. Some lie within our muscles, telling us where our body is and how it's moving; or within the vestibular organs in our inner ears, keeping us balanced physically if not psychologically. Some are buried in our internal organs, monitoring their function so that if necessary the brain can shift resources from one area to another: from the guts to the muscles, for instance (in an emergency efficient movement takes priority over efficient digestion). The contents of the blood, the behaviour of the immune system, levels of chemicals in body tissue, gut contents: the brain processes information about all of these.

We tend to become aware of this only *in extremis*, when toxins, for instance from badly-prepared food, get into our bloodstream and are detected by the brain. This is not as simple as it sounds, because brains have sophisticated border controls: the blood–brain barrier, which protects our vulnerable and valued property from chemical attack. Like any security system, the blood–brain barrier has to balance security with the need for information. Hazards must be kept away from delicate brain cells; yet the brain must be informed of their existence in order to trigger appropriate body defences. Vomiting and diarrhoea, for instance, help to get rid of the stuff in your gut which may (or may not, but better safe than sorry) have been responsible for the toxins. The immune system triggers the alarm and does much of the groundwork, but the brain's essential contribution implies that the blood–brain barrier cannot be too efficient at keeping things out.

Natural selection has resolved the trade-off by developing specialized areas like the area postrema, a segment of neural tissue which extends outside the blood–brain barrier, allowing it to monitor the contents of your bloodstream, detect potentially harmful changes, and respond with appropriate gastrointestinal consequences—without fatally compromising brain security. Once activated, the area postrema sends signals to areas at the base of the brain (the medulla) which have a direct connection to the stomach. The circuit is beautifully efficient, quickly removing the potential source of the contamination without the need for lengthy consultations with senior management (although reports may be sent to areas of cortex such as the insula, which processes signals from the viscera). The area postrema is a stunning example of evolutionary problem-solving. All we know, however, is that we throw up.[2]

In addition to signals from our innards, the brain is capable of generating its own internal alerts in response to external signals. Specialized clumps of cells (nuclei) in the base of the brain send their filaments through the cortex and subcortex. When activated, this 'vigilance system' spreads neurotransmitters such as noradrenaline (a close relative of the world's most famous stress hormone) through the brain. This makes neurons trigger-happy—much more sensitive to incoming signals than usual, and thus better at detecting that tiny rustle of bushes or movement in the shadows which might signal an imminent attack.[3] Unfortunately, in today's crowded world the threats are new and numerous, the predators more subtle and more various, and many of us feel that we live with chronically hyperactive vigilance systems.

Vigilance alerts have much in common with signals from the body's viscera. Both originate internally, have wide-ranging and non-specific effects which tend to linger (i.e. they shake up the whole brain, not just a few areas), and do not require an external stimulus to trigger them (human beings can work themselves into paroxysms of anxiety over non-existent, merely possible events). These internal signals are the neural events which we interpret as emotions, moods, or feelings (which I will lump together as 'emotions', with apologies to purists). When we face the choice of whether or not to do harm, emotions can weight our reasons to act so heavily that common sense and even self-preservation stand no chance against them. Questioned later, we often cite our strong emotions as causes: 'I struck first because I was frightened.' Yet emotions are not as simple as this suggests. They can be as clear as lightning, and as overwhelmingly powerful—but they can also be so slippery and ill-defined that a person may say: 'I have no idea why I did that,' and not be lying.

One baffling aspect of emotions is their variability and the variety of their causes. Worldly or bodily, real or unreal events can move us, and through multiple senses or imagination. A person may be disgusted by the sensation of swallowing an oyster, the sight of one, the sound of one being slurped up, or just the idea of choking down that gloopy little mass. The same person, confronted with a challenge, may feel overwhelmingly angry, afraid, excited, or all three and more. How often, in fiction or real life, do we encounter mixed or indefinable feelings, emotional turmoil, powerful urges, and the like? How can emotions be both body-graspingly potent and inarticulately nebulous? To approach these questions, we need to know more about what emotions are and what they are for.

What are emotions?

Emotions are found in every known human culture. Their ramifications can enrich, or wreck, our lives; without them we would be scarily stunted creatures. Yet many of our most prominent intellects have regarded emotions with roughly the level of appreciation the rest of us reserve for estate agents and dirty laundry. Humanity was christened *Homo sapiens*, after all, reflecting our status as rational animals, with reason defined in strictly cognitive terms. 'Je pense donc je suis', proclaimed the philosopher Réné Descartes: 'I am thinking, therefore I exist.'[4] This glorification of dry reason culminated, in the early twentieth century, in a view of reality which saw abstract cognition (i.e. logic and mathematics) as pre-eminent, and mathematicians, scientists, philosophers, and chess-players as the highest form of life (perhaps unsurprisingly, since they were its chief promulgators).[5] Emotion was dirty, unreasonable, female (an ancient linkage, that one), sunk in the body, and no help at all when it came to proving theorems.

It would be nice to record that this 'logicalist' world-view faded from fashion because its senior thinkers saw the error of their ways. To be fair, some of them did.[6] In reality, however, the paradigm shift probably had more to do with logicalism being a victim of its own success. The application of all that dry reason led to the production of weapons so spectacularly abhorrent that even the driest reason might flinch at their consequences. It also created computers so proficient at chess and mathematics that human supremacy in these domains began to look distinctly shaky. By the time a machine finally beat the chess world champion, Garry Kasparov, under tournament conditions (in 1997), the advantages of incorporating emotion in the human definition were clear. Machines might topple grandmasters, but they could be outfelt by the feeblest infant. The logicalist habits of exalting rationality, regarding clever young white men as representative of the human species and downplaying the fact that brains are located in bodies, began to be recognized as somewhat problematic. These habits are still a strong presence in neuroscience and psychology, as elsewhere, but in recent years they have been joined by new approaches. Studies of intuition and cognitive biases, social and cross-cultural psychology, and affective neuroscience are all fast-growing fields.

Modern affective neuroscience is very young. Arguments rage about how to define emotions—and even about such apparent basics as how many there are and whether you can tell them apart.[7] This may have something to do with academics' love of debate, but it also reflects the slippery nature of the subject-matter. What is clear is that emotions have much to do with states of the body. Physiological changes give us the tremors of fear, the red face of fury, the nausea of disgust. There is also wide though not universal agreement that some emotions (fear, anger, disgust, sadness, and joy are often cited) seem to be simpler and more basic than others (like shame, embarrassment, the Greek *hubris*, and the German *Schadenfreude*). Basic emotions have characteristic facial expressions and physiological responses, so characteristic that every culture seems to know them; an angry Aztec or disgusted Diego Garcian could easily communicate their feelings to you or me.[8] Yet even basic emotions can be confused: anger with fear or excitement, for example. Moreover, many of the emotions we feel are pale shadows of the full experience. We often use emotional language without feeling anything intense—as when we say jokingly, 'I hate that', implying only casual aversion. And sometimes we can't say what it is we are feeling.

Part of the reason for this inarticulacy lies in the nature of the signals of body state from which emotions arise. Although these reach the brain and influence neuronal activity, just as external signals do, there are differences. Visceral messages from the gut, for instance, seem to be slower and less precise than their somatosensory ('body-sensing') counterparts.[9] Internal signals tend to last longer than those created by rapidly changing sounds or images; this is particularly true of signals carried brainwards not by nerves but by hormonal (endocrine), immune, or other chemical messengers. Prolonged in time, internal signals are also less clearly defined in space than their visual or auditory counterparts, so their influence may be hard to assess.[10] Neurons which process sounds, for instance, allow the brain to distinguish the subtle differences in frequency and pitch which underlie our capacity for language. Function is crucial: the 'use it or lose it' soundbite seems to apply to brain tissue, down to the level of individual synapses. Frequently used muscles, like those of the vocal cords, mouth, and hands, commandeer large neural resources in both sensory and motor areas of the brain, allowing the nuanced movements and detailed discriminations necessary for communication by word, grimace, or gesture. Neurons in less well resourced domains, such as those which process sensations of anal extension, do not need to make such fine distinctions, as *Homo sapiens'* ability to talk through its arse is strictly metaphorical.[11]

Emotions arise from brain signals caused by changes in body state (what the neurologist Antonio Damasio calls 'somatic markers').[12] Those signals can sometimes be strong—demands that the body's happenings be attended to—but they are not necessarily precise, in the way that information about where your left hand has just moved is precise. Yet we, or at least our finest writers and artists, can differentiate innumerable shades of mood and meaning. Smiles, for instance, can be sweet or acid, open or secretive, fixed, friendly, or furtive, and much more.

EMOTIONS AND LANGUAGE

To achieve this subtlety the body's non-linguistic messages must be set in a symbolic frame. Like an actor's grunts and groans, they need context to be understood. However, we cannot simply say 'to make an emotion, take one body signal, add language, and hey presto', because, inevitably, it's more complicated than that. Let us take the acting analogy a little further and imagine watching a movie scene in which a man is being hit and groaning. How do we make sense of his behaviour? By using our own experience and our knowledge of his context. We may never have been beaten up ourselves, but we see a man being hit and wince in sympathy, interpreting his noises as pain-behaviour. Simultaneously we are keeping track of plot developments and characters' relationships, not to mention the terse dialogue, rapid set changes, and clever cultural references. We probably haven't time to say to ourselves 'that man's grunting like that because he's in pain', and nor do we need to frame our understanding in language. Like dogs or chimpanzees, we grasp much about others' emotions without needing words.

Sometimes, however, we need to communicate feelings. Easy: we just imitate the relevant behaviour. But what if we are in a situation where grunting or weeping or punching the nearest onlooker isn't advisable— at a garden party, for instance? What if we are trying to write a novel, whose readers will not be able to see us acting out the dramas? Now we must resort to language and give our feelings names.

Language lets us associate specific symbols (words) with the objects and events we perceive around us, their relationships and the ways in which we interact with them. We learn language in the way we learn anything else, by detecting correlations—changes in the environment, our bodies, and our brains which seem to go together. Sometimes the link between an object or event and the sound, sight, or thought of the word which refers to it is obvious. An infant can compare the sound his mother makes at the sudden appearance of a ball—'Ball!'—with the sounds she makes when he picks up a ball—'Ball! Clever boy!'—and the way she greets the sudden appearance of a dog—'Doggy!'. Over many repetitions, and given all the other, non-verbal clues that mothers provide, the sound 'ball' correlates better (goes together more often) with round objects fit for throwing than it does with, say, objects which bark or have legs.[13]

The subset of language which uses theory of mind extends this useful facility into our heads, to the 'bits in between' perception and action. Here the neural patterns generated by our sensory awareness merge with those reflecting our past experience, current expectations, and visceral concerns, like waves from many rivers pouring into the same hugely turbulent sea. Even in this maelstrom, however, there is order. Some distinctive patterns reproducibly correlate with sensory or motor constructs and reflect relationships between them, like the relationship between packets of cigarettes and mortality implied by 'smoking kills'. If these in-between patterns are distinctive and stable enough, they too can be associated with words. To describe them, to ourselves or others, we use the language of cognition: knowledge, beliefs, ideas, hypotheses, and so on.[14]

Other 'in-between' patterns, however, seem to reflect our state of being rather than the world around us. They don't correlate particularly well with objects, on the whole, but they do go together with bodily sensations, from the galloping heartbeat of fear to the muscular relaxation of happiness. We often associate them with a compulsive quality, motivation, and choose our actions so as to reduce or increase them. These patterns may not be as distinctive as cognitions or perceptions, but they do vary in ways which are stable enough for us to detect and link a name to, depending on how our body and brain react to what we perceive. The names our culture has chosen—anger, disgust, contentment, guilt, and many more—can then be generalized to other situations in which what we feel may be less intense, but similar in type.

Emotion language differs from learning how to talk about objects in that one cannot simply point out what the words refer to. The day may yet come when children are taught emotion words with the aid of fMRI scans, but for now parents and care-givers rely on strategies which make the words and their correlations more distinctive, more salient to the learning child. Interact with infants, and you will almost certainly find yourself doing the same: using techniques like vocal emphasis ('Baby *likes* that!'), differential definition ('She's *not* happy! She's *sad*!'), exaggerated non-verbal cues (accompanying 'Yuck! That's disgusting!' with clownishly extreme disgust-expressions), and so on.[15] We also take a huge amount from social learning. Brides are expected to feel happy; the bereaved, sad; people given approval by a superior, proud; new recruits, nervous; and so on.

Once the core symbols have been established we can weave them into our linguistic networks, associating them with the perceptions, cognitions, and behaviours in whose company they seem to spend most time. We can define them in great detail or leave them largely unexamined, depending on personality, intelligence, and the demands of experience. Cultural as well as individual differences come into play here. Some societies may emphasize particular nuances, while others do not (or simply pinch their neighbour's vocab, as English-speakers did with *Schadenfreude*).

This process of interpretation, which imposes some order on the neural chaos, changes the patterns as it makes cognitive sense of them.[16] The brain activity bridging the gap between a registered insult and the clenching of fists may not be labelled at all, or it may be interpreted as rage, injured pride, dismay, or the eagerness to fight. Moreover, these emotion-labels are not simply there for decoration or to tidy up the mental landscape. They matter, since the person's decision about what to do next may be affected by how he or she decides he or she is feeling.

In short, emotions may begin with signals triggered by changes in body state. As they reach the brain, however, those signals flow into cortical networks set up by experience, influencing brain activity in ways we may or may not be consciously aware of. That activity, which will often involve sensory and motor components (e.g. facial expressions) as well as cognitive and visceral inputs, may then be incorporated into the symbolic world of mind and language by silent self-analysis or public expression. By 'emotion', therefore, we can mean any or all of these sensory-motor-visceral-cognitive-experiential 'bits in between'.

What are emotions for?

Body state matters. It is easier to tune out a barking dog than to ignore a migraine. Visceral signals appear to dominate brain activity more than their sensory equivalents.[17] They are like especially loud-voiced members of our neural committees, demanding that something must be done. Disregarded, these voices grow more and more strident until they can no longer be ignored, and we change our behaviour in order to make them happen less often. That at least is the case for aversive emotions, to which we have evolved to pay more attention because of their tendency to warn of trouble. More positive feelings reinforce, rather than inhibit, the behaviour which brought them about.[18]

With experience we learn what needs to be done, and emotions can push us into doing it faster. As noted earlier, we can think of them as the memories of decisions made by our genetic predecessors (for evolved reactions like fear) or our former self (for more complex emotions)—personalised potted histories of relationships with the stimulus in question. Or, more precisely, records of the consequences of decisions. Like guidance from ancestor spirits, emotions hint at what will serve us well, or badly, based on experience: 'Give her a hug, you'll feel better'; 'Don't eat that, you'll regret it'; 'Want this opened? Do what you just did, but harder.' Coupled with language, we can add our own time- and effort-saving messages: 'Avoid the boss on Monday mornings'; 'That's my favourite'; 'People with shifty eyes are not to be trusted.'

Emotions are thus 'for' informing us about our body, as the term 'somatic marker' might suggest.[19] They also tell us much more: about other people's feelings, our own evaluations of objects we encounter and social interactions, the events which flow from what we do and those which prompt us to act. Pure fear, pure joy, pure anger—these require a background before we can make sense of them. Body signals like the 'rush' due to adrenaline affect our physiology in much the same way whether the cause is a real threat, an imaginary hazard, or an injection of the hormone into our blood. Yet the adrenaline rush may be felt as terror, stress, or thrilled excitement depending on the person and their circumstances.[20] What all three interpretations share, however, is their ability to highlight certain events as salient, making them stand out from the neural crowd and prioritizing responses to them. Emotions thus confer meaning in the sense of labelling certain neural patterns as important to the person. These things matter to me; they affect my behaviour. I will make more effort to avoid them, or to pursue them.

Negative emotions in particular are also warnings, prioritizing rapid reactions to threats. I previously described three common response programmes which allow humans (and other species) to respond quickly and effectively to familiar categories of threat. Escape-oriented behaviour—which includes submission and distress signals, freezing, crouching, or fleeing—evolved to prevent human aggression from escalating and to avoid being noticed or caught by predators. The emotion-names we use in such situations include fear, fright, terror, and the like. Challenge-oriented behaviour—which involves threat signals such as glaring and bared teeth, sometimes with physical aggression—serves as a display of power, and is directed against other agents, or objects interpreted as agents (as in 'My [expletive] computer's crashed again!'). Here the emotion-names cluster around the concepts of anger, pride, and outrage. Avoidance behaviour, which includes withdrawal, gaze aversion, self-cleansing, spitting, nausea, and vomiting, evolved to protect humans from poisons and contaminants—things which weren't agents and weren't sufficiently powerful or imminent to trigger escape behaviour, and yet could be highly dangerous if approached.[21] The emotions involved in such cases we class as disgust, horror, and their kin.

To understand how emotions are constructed by the brain, we need a specific example. I will focus on disgust, not from innate obstreperousness but because disgust, one of the most distinctive and yet least-considered emotions, has a crucial role in otherization and cruelty.[22] We have already come across one of its components: the vomiting reflex mediated by the area postrema. We have also noted that disgust has much to do with cruelty. So how do human brains achieve disgust?

Constructing disgust

> *car la vie, c'est la pourriture* [for life is corruption]
>
> (Jacques Lacan, *Séminaire VII*)

Disgusting stimuli trigger a suite of reactions in the perceiver, including changes in heart rate, breathing, and brain activity, a characteristic facial expression (wrinkled nose, pursed lips, narrowed eyes), and nausea or vomiting.[23] Queasy aversion is the core response, culminating in the gag reflex which ejects the stomach contents. Overlaid on this basic function are layers of increasingly sophisticated behaviours which allow for early detection and avoidance of disgust-threats. Not eating the rotten meat in the first place is safer and uses less energy than eating it and having to throw it up again. Disgust in humans, however, is far more complex than mere programmed avoidance. It is at least partially under our conscious control.

That disgust involves voluntary control is clear. Thinking about disgust, deliberately making disgusted facial expressions, or even reading the word 'vomit' can give rise to physiological reactions as well as to the subjective emotional experience of disgust.[24] Voluntary control makes evolutionary sense for social beings: emotion regulation is a critical social skill. It is not always politic to show disgust; nor is it always possible to employ the three usual tactics of avoiding the disgusting object, expelling it from one's presence or eliminating it by some other means. Moreover, disgust, like other basic emotions, has many nuances—distaste, contempt, abhorrence, abomination, revulsion, and so on—suggesting the importance of conscious, language-based inter-pretations in defining this emotion, and its crucial social role. Thus the neural mechanisms which generate and express disgust must be layered, allowing both for automatic protective reactions (like vomit-ing) and for high-level conscious control of emotional responses, mod-ulated by awareness of the situation and the assessment of alternative behaviours.

THE NEUROANATOMY OF DISGUST

In the sea of unknowns that is neuroscience the brainstem circuits involved in vomiting constitute a small island of understanding. The layout is as follows. When nerves in the stomach detect the presence of something nasty they send signals along the great sensory highway of the tenth cranial nerve, the vagus, named from the Latin word for wanderer.[25] This cranial vagrant delivers the stomach's alarm to that skilful segment of brainstem, the area postrema, which in turn contacts other areas in the pons and medulla, the base of the brain.[26] These com-mand the stomach muscles, triggering the rhythmic contractions which we feel first as stomach-churning unease, then nausea, then gagging surrender.[27]

Nerves from the stomach also reach another brainstem area, the nucleus tractus solitarius (NTS), which has close links with the area postrema (and with other bits of subcortex involved in managing innards, like the hypothalamus and amygdala).[28] The NTS contains var-ious clumps of cells with different priorities: some deal with swallowing and sensations from the larynx, pharynx, and stomach, some with breath-ing, others with blood pressure. All need adjustment when vomiting takes place, so that the airways are closed for long enough but not too long, the stomach wall efficiently contracted at the right time, and so on.[29] Considering the organization required to synchronize all this, one can only be grateful that some brain functions come pre-supplied; having to throw up is quite bad enough without having to think about how to do it.

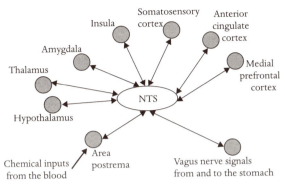

FIGURE 7. Some of the connections involved in the brain's processing of disgust. Chemical changes in the blood are detected by the area postrema, which sends signals to the nucleus tractus solitarius (NTS). The NTS also receives signals from the stomach, via the vagus, and sends the signals which trigger vomiting. In addition, the NTS is connected to subcortical areas like the thalamus, hypothalamus and amygdala, to the insular cortex, and to other regions such as the cingulate, primary somatosensory and medial prefrontal cortices (all of which are also interconnected).

Given a strong enough trigger, you will be sick. For lesser stimuli, however, signals from other parts of the brain can facilitate or inhibit the gag reflex. Connections from areas such as the somatosensory cortex (see Fig. 5), insular cortex, anterior cingulate cortex, and medial prefrontal cortex (all of which also talk to each other) reach down into subcortex, allowing them to suppress or enhance the extent of the disgust experience (see Figs. 7–9).[30] Cortical areas, particularly the insula, also receive inputs from the brainstem control centres, keeping them in touch with the visceral changes. The more areas that become involved, the 'richer' the overall experience will be.

In addition, once the networks of neural connections are established by experience, activating part of a network will tend to lead to the rapid spread of activation across associated networks (strongly connected synapses), 'filling in' the neural pattern (as discussed in Chapter 4). This is why humans do not need to have toxins in their body to feel sick. Pictures of car crashes, sounds of someone vomiting, the feel of decaying meat, foul smells or tastes, even purely symbolic disgust-threats can all achieve the same effect by activating similar cortical and subcortical areas (for example, networks of visual cortical neurons which have synaptic links

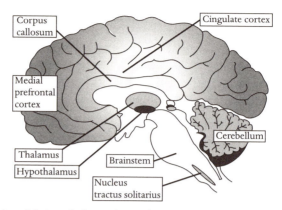

FIGURE 8. A medial view of a human brain. Three major landmarks are labelled: the cerebellum, brainstem and corpus callosum (the large band of fibres which links the brain's two hemispheres). Also shown are the cingulate and medial prefrontal cortices, the thalamus and the hypothalamus.

with the insular cortex and amygdala): those which participate in what we call disgust responses.

Once we climb from brainstem to cortex, however, we have left the island once more for the open ocean. Scientists have only recently begun

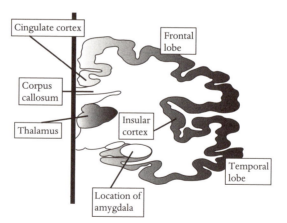

FIGURE 9. A coronal view of half of a human brain (as if the brain has been sliced vertically through its front half). The cortical grey matter is shaded and two landmarks are labelled: the thalamus and corpus callosum, along with the temporal and frontal lobes and the cingulate. The location of the amygdala, which lies buried deep inside the temporal lobe, is indicated. This view of the brain also shows the location of the insular cortex, hidden beneath the folds of the temporal and frontal lobes.

to untangle the cortical processing of disgust.[31] The insula, for example, has been linked to violent behaviour, such as reactive aggression and conduct disorder, and even to the process of rejecting beliefs.[32] Clearly, much remains to be discovered about the roles and relationships of this and other areas of cortex.

Emotions and 'shadow' emotions

Emotions frame connections between stimuli and behaviour in terms of the impact of both upon the body. They also strengthen these connections, allowing emotive events to be better remembered and more influential in future decisions to act. Stronger connections, as noted in the previous chapter, elicit faster responses. The full repertoire of an emotional expression, however, takes time to unfold—facial muscles must be rearranged, hormones released, breathing patterns altered, guts churned, and so on. Consequently, as responses get swifter they may come to be triggered before the emotion is fully felt. As the emotion-processing networks involved are increasingly bypassed by stronger, more direct input–output pathways they lose intensity, becoming shadows of their former selves.

The desensitization which hardens multiple murderers into callousness relies on this emotional habituation.[33] It is particularly effective when used along with stressful demands to act quickly, ensuring that the person literally does not have time to feel. Even in mild and trivial circumstances, mere repetition can greatly reduce emotional power. As an example, consider that favourite standby of moral philosophy, the trolley problem.[34] As I first encountered it, this asks you to imagine a trolley track along which five people are happily rambling. There is not enough room on either side for them to get off the track, and they do not know—but will soon find out—that a trolley has broken loose and is hurtling towards them at lethal speed. You are standing beside the line further up. You can either do nothing, which will leave five people dead, or you can move a lever which will divert the runaway vehicle onto a branch track, killing one person who happens to be walking along it. The choice is yours. Needless to say, most people choose to move the lever. Causing any death is wrong, but when it comes to setting one stranger's life against five, utilitarian cost–benefit analysis governs the calculations.

At this point, you discover that there is more: a second version, identical to the first except for one crucial change. You are now standing on a bridge above the track, and beside you is a very large man. Your body would not be big enough to block the line, but his would. Just one push, and you have saved five people's lives. Will you do it? In terms of death and rescue the tasks are identical, but morally they are quite different: most people react with horror to the second version. This brings joy to philosophers and scientists, who can then spend ages investigating our illogical moral judgements.

I first encountered the trolley problem so long ago that I have no idea how I reacted to it (it was probably the way most people react: flick the switch, don't push the fat man). Repetition, however, has led to desensitization, and the irritated conclusion that the best solution might well be to do nothing in both cases. Trying to shove a fat man to his doom sounds like lunacy (he's bigger and stronger than me; why wouldn't he just shove me over?). Besides, letting five people die will make a much bigger media splash than killing one, thus providing a more effective warning to any other idiots who plan to go walking on tracks where they can't get out of the way of oncoming vehicles.[35] Fortunately for careless ramblers, the trolley problem is hypothetical and my hyper-utilitarian callousness consequently immaterial. Faced with a real moral crisis I would probably opt for the usual response. Real moral crises, however, are equally subject to habituation. Even killing children, let alone pushing fat people off bridges, can become routine if you do it often enough.

Shadow emotions involve cortical layers of emotion circuitry without reaching down through the brainstem to evoke the full physiological response. They can be triggered by highly abstract stimuli—words or thoughts, whether spoken, written, or remembered, or even the imagined visions of a daydream—sometimes so briefly that the person is unaware of how their mental landscape has been tinged.[36] A wealth of research, however, suggests that even fleeting emotions, like those aroused by loaded language, can bias responses, enhancing emotion-congruent thoughts and behaviours (those which have become associated with the emotion in question) while suppressing emotion-incongruent reactions.[37] This ability to link emotions to purely symbolic threats lies at the heart of otherization. As we shall see in Chapters 7 and 8, strong emotions have much to do with callous and sadistic cruelty—but shadow emotions can also play a part.

Emotions and cruelty: a case study

As an illustration of the arguments set out so far, let us consider the following brief story about a young child's encounter with her family's pet. First of all let's have the facts:

> One day, the child approached the cat with a piece of string tied into a loop. She slipped the string over the cat's head and around its neck. Feeling the string tighten, the cat struggled. Seeing what was happening, the child's mother came running, shouted at the child, and freed the cat. Mother and child then talked about what had happened.

What is your reaction to this story? Would you feel shocked if you saw a child treat an animal in this way? If so, were you more shocked because the child was a girl than you would have been if a boy had been involved? Or were you amused (perhaps the scene reminded you of cartoons you watched as a child)? Did the thought of the cat's experience fill you with pity, so that you may even have felt it as painful, a tightness in the throat or a lurch in the stomach? Maybe you strongly dislike cats and wish the child had succeeded in killing this one. Maybe you think no one, not even a child, should get away without severe punishment for tormenting an animal. Or maybe you feel nothing much at all.

Whatever your reaction, it probably involved using theory of mind to analyse the child's behaviour in terms of her beliefs and motivations.[38] The story deliberately provided no information about these. Yet we read it and wonder: why did the child act as she did? What did she think would happen? What did she want to achieve? In doing so, we make guesses about her motives which are crucial to how we evaluate her behaviour. The trauma inflicted on the poor half-strangled cat remains the same, but 'childish ignorance' and 'shocking cruelty' have very different consequences for our moral judgement of the perpetrator.

And moral judgements, especially of children, can have a massive impact on the lives of those assessed. We treat our moral judgements about a person as predictive of his or her future behaviour, and shape our own accordingly. Compared with one labelled merely 'ignorant', a child labelled 'shockingly cruel' is more likely to face a bleak future; in the popular imagination childish cruelty is linked to adult venom.[39] We fall easily into the essence trap, assuming that while ignorance can be corrected, cruelty is an innate and irredeemable flaw of a person's nature. Natures being much like souls (that is, either difficult or impossible to change, depending on your preferred flavour of biology and/or theology), the temptation may arise to take a shortcut and remove the behaviour by removing the person. Those we call 'cruel' we find much easier to ostracize, lock up, or kill.

Moral judgements are not necessarily fixed from the moment they are formed. As we gain new information, we often revise our original guesses. If you have selected the 'cruelty' explanation, and I tell you that at the time, the child appeared highly distressed by what had happened, then you may switch to the 'childish ignorance' interpretation instead. Then again, a budding psychopath might have had the cunning to burst into tears at the appropriate moment, and many mothers have been fooled by psychopathic offspring. What, then, if you learn that the child matured into a relatively law-abiding adult who is careful to avoid killing even spiders, let alone anything mammalian? This comment on the child's future *behaviour* does not refer directly to her *feelings*. Nevertheless, it may convince you that she acted out of ignorance, because her behaviour over time is consistent with her being concerned for animal welfare. Cruel people, in our experience, do not waste time and energy removing spiders from the carpet to the great outdoors. Why would they? Rights for spiders are hardly a high priority even for animal-welfare organizations, so a sadistic spider-squasher need have little fear of retribution.

Reasoning about other people's actions looks for the most obvious explanation, given what information is available. That information includes knowledge of the people themselves, as well as a huge amount of commonly shared social knowledge, built up in each of us over time as a result of experience. For example, we know that in general people put a lot of effort into something only when they have a very good reason for doing so. We cannot see an obvious reason why the child, following her encounter with the cat, should spend many years of her life being kind to animals—and particularly invertebrates, who have few defenders—if she secretly wishes to hurt them. We can, however, see an obvious causal connection between initial genuine distress and later concern to avoid hurting animals.

Let's dig a little deeper:

> The mother told the child that she had hurt and frightened the cat, and could have done it serious harm. She asked the child why she had done this. The child, in tears, explained that she had wanted to take the cat for a walk. She had seen dogs being taken for walks on leads, so why not cats? But she didn't have a lead, so she had made one.

This new information shifts our judgement towards ignorance and away from deliberate malice. Once again, though, we can and do extrapolate well beyond what we've been told. It is easy to imagine and perhaps empathize with a child who has grown up in houses with cats, whose parents are tender towards their cherished pets. We can understand what it felt like not to mean to hurt the cat. We can grasp that the child knew enough to see that dogs and cats were similar in many ways; to be able to make a piece of string look—and to some extent function—like a lead; and to recognize immediately the signs of the animal's distress. We can accept that, knowing so much, she might nevertheless not realize that cats (and even dogs) need training before they will tolerate a lead; nor understand that, without a carefully placed preventative knot, a loop of string around the neck of a resisting animal will keep on tightening. All this requires dipping into a sea of background knowledge, which we do effortlessly, but a child may not yet have learned what we take for granted.

WHAT DOES OUR EXAMPLE TEACH US?

Several points arise from this little fable. First, context is hugely important. Did you accept the child's sincerity? If you did, one element of the story which may have helped to persuade you was the child's mother's reaction after the event. Her initial anger fits a template familiar to us all from childhood fairytales, TV soap plots, and tabloid headlines: Person A (good) comes across Person B (bad) doing something wicked; Person A reacts with outraged moral condemnation. But her next action was to talk over what had happened with the child, explaining the hurt she had caused—not simply meting out punishment. This does not fit the pattern: bad people are usually condemned, demonized, punished, or introduced to a suitably bad end. The mother can be expected to understand her child; we can see her, therefore, as a trusted authority when it comes to assessing the child's motives. If she acts as if she believes the child, perhaps we are justified in doing the same.

Secondly, our interpretation is a matter of reducing the number of possible explanations. The initial story provided no reasons why the child acted as she did, but in practice we focus on relatively few of the infinity of possible stories. These serious contenders may include mental illness, cruelty, ignorance, hunger, curiosity, the belief that cats are vermin which ought to be killed, or someone ordering the child to kill the cat. These are explanations we frequently come across in situations where someone has hurt or killed another sentient being. Experience tells us that some are more likely than others (most children we know are unlikely to be that hungry). We judge their likelihood on the basis of 'common knowledge', which in practice includes common prejudices, stereotypes, and unjustified biases. For instance, the mother's decision to talk to her child rather than punish her may make the 'mental illness' and 'cruelty' explanations appear less likely. The relevant stereotypes hold that we don't talk to the mentally ill (unless we are professionals whose job it is to treat them), and we punish cruel people (or else risk condoning their behaviour).

Each additional piece of information helps us narrow the field still further, so that we are quickly able to reach a conclusion about the child's motives and beliefs. This point is perhaps worth reiterating. Although in principle a living brain faces an infinity of potential responses to any given situation, in practice that maze of possibilities is highly constrained, narrowed to only a few potential responses by the time the responses themselves are due to be triggered. Agonies of indecision generally involve two or three options, not hundreds or thousands.

Thirdly, emotions have much to do with morality. How we feel about the child affects how we judge her. Recent psychological experiments, for instance, show that manipulating people's feelings of disgust without their knowledge affects the moral judgements they make, their willingness to help others, and so on.[40] Not realizing the causes of their disgust, thanks to canny experimental design, they seem to attribute the emotion to the task (e.g. making a moral judgement). The more disgusted we feel about a person or behaviour, the more harshly we appear to condemn them.[41]

Finally, what happened to the child illustrates the slippery nature of emotions. To understand this, we need to know what she was feeling at the time. Clearly, this is not fully possible: we are not in her circumstances, with her background, genes, and so on. Nor does language provide an exact description of emotions, despite the many nuances available. However, we can construct a plausible narrative of her inner experience, as follows.

The child's first reaction was shock. She expected a tamely strolling cat and had never seen the family pet behave in such a way. Her shock involved a high degree of what psychologists call 'arousal'— hyper-alert attentiveness, an adrenaline rush complete with racing heart, muscular tension, and suchlike—but with no sense of accompanying positive or negative emotion. It was like being balanced on the sharp peak of a mountain, knowing that one would fall one way or the other, but not yet knowing which way. The child, unused to the feeling, simply froze, staring at the cat, unable to interpret what she was experiencing.

Then the child's mother arrived. Her behaviour gave the child the cues she needed to select one particular emotional interpretation—in her case, a strongly negative one—of what she was feeling. Her mother's anger was experienced as painful, not as stimulating or provocative, and the rush of adrenaline was interpreted as fear and dismay, rather than as excitement or euphoria. That interpretation was reinforced by the conversation that followed.

If the child's mother had scolded or beaten her, had ignored her behaviour, or had simply laughed indulgently, might that child's interpretation of her feelings, and her consequent behaviour, have been different?

One must not exaggerate. For this particular child, lucky enough to be born into a loving middle-class environment, that particular encounter is not likely to endow her with anything other than anxiety, social compliance, and concern for animals. Her parents abhor cruelty; their friends and relatives consistently approve of kindness and good behaviour; her culture exposes her to images of animals as 'friendly' and human-like (e.g. in Disney films), and so on. There is no single 'tipping point' for this child; just the repeated message that 'hurting animals hurts them and will hurt you', whereas treating them with tenderness feels good and brings rewards.

Other children get different cues to guide them. The messages in their world lead them to interpret what they feel—internal signals arising from their body and brain—in positive terms, associating their behaviour with excitement (i.e. reward) rather than anxiety and stress (i.e. punishment). Hence the importance of supplying immediate and consistently negative messages about the *behaviour* (not about the *child*; that falls into the essence trap, and changing behaviour is much easier for the child to understand than changing the self). Research suggests that much child cruelty to animals (and, indeed, to other children) occurs in group situations, where the participants encourage each other's harm-doing. Having said which, we must not exaggerate here either; cruelty in childhood is not a guarantee of future monstrosity. Many of today's respectable adults will have done dreadful things to a living being on occasion, not just accidentally but with full intent (like the bishop I once saw at a picnic who used his fork to mash an innocent wasp). Most such offences either do not count, in our cultural lexicon, as cruelty (can wasps suffer?) or are unpleasant rarities, shaming memories which encourage their holders to be nicer, kinder, gentler types in future.

Summary and conclusions

Signals from our bodies provide us with the information we need to understand the impact of events—their importance for us. Our external sensory systems tell us what is happening; our motor systems tell us what we are doing about it; to complete the loop, we need to grasp the consequences of our actions and of changes in the world around us. It is not enough to know that we are eating, or even that the food tastes sweet; we need to know what it does to us in order to know whether to eat that food again or to avoid it next time round. Does it lift our mood, send us to sleep, make us feel queasy or bloated, give us a buzz? Do we like it or do we not? How does it make us feel? This is the realm of evaluation, where emotions hold sway.

Evaluating an event or interaction as good or bad, in the simplest cases at least, depends on the physical effect it has on us. A hot stove burns Joe? He'll be more careful in future. A conman cheats Miriam? Sickened by her losses, she swears not to fall for that scam again. Events are more real to us if they affect us directly, causing us pain or giving us pleasure. They matter more and have more meaning for us; in neural terms, they produce more distinctive activity patterns which are more likely to alter our future behaviour. Eating a food which actually causes you to vomit will be far more likely to make you avoid it in future than eating a food which you know made someone else sick, or reading about a meal which made somebody ill. Yet even the briefest allusion to something disgusting, like mentioning germs, rats, or tumours 'in passing' when talking of Jews, can conjure the patterns of disgust to some extent.

The technique of creating these negative, otherizing emotional links without their habituating into blandness involves repeating the verbal association frequently, but not too frequently, and casually (throwaway remarks look more sincere) rather than with enormous emphasis. This allows the link to strengthen without arousing conflicting beliefs ('don't be ridiculous; people are nothing like tumours') which might lead the person to challenge and weaken it. Like all otherizing techniques, it is simple and intuitive, exploiting the brain's natural penchant for making connections. Every demagogue and genocidal leader in history has mastered it, and no atrocity would occur without it.

Emotions provide some of the reasons which lead people into cruel behaviour. A motivation to act, however, is in itself imprecise and insufficient: to be effective, its force must be channelled into triggering specific behaviours. The channels used by human beings are none other than beliefs. Accordingly, we turn next to these strange denizens of the mind: what they are, how they change, and how they can make us cruel.

Chapter 6

How do we come to believe?

State-perpetrated or -tolerated physical violence towards an identifiable group could not occur unless it is preceded by symbolic violence.

(Carole Nagengast, in A. L. Hinton's *Annihilating Difference*)

What are beliefs?

At first sight, as we leave behind the turbulence of emotions, the topic of beliefs seems as reassuringly solid as a flight of steps, taking us up to the dry, clear world of cognition. Beliefs, after all, are information-bearing entities, with all the scientific and logical connotations that term implies. Linked by bonds of reason, they form the vast networks of knowledge from which we assemble our identities and through which we interact with the world. Like statements, they can be grammatical or ungrammatical, meaningful or nonsensical, and we have sophisticated capacities for telling which is which. They can also be true or false, and if we find them to be false we can correct or discard them.

This allows us to sift and winnow our beliefs—either by using logic to test them against each other or by gathering evidence to test them against reality—and (by keeping only the true ones) to maintain an accurate understanding of how things are. When the world around us changes, our beliefs change to match. As long as our abilities to reason and to perceive are unimpaired we can therefore assume that most of our beliefs are true. The more strongly we believe them, the more certain we can be that they are true. We assume, in other words, that however hormones may surge, moods swing, and feelings ebb and flow, we can trust our beliefs to be securely rooted in reality.

REASONS AND EVIDENCE

It's a nice picture, and it's not altogether wrong. Some of our beliefs do derive their truth from logic; that is to say, from their symbolic relations with other beliefs. Take the following three examples:

- $120 + 4 = 124$;
- colourless ideas cannot simultaneously be green;
- if all enemies want to kill you and Fred is an enemy then Fred wants to kill you

These convictions are inescapable once you understand how the symbols involved are to be used (the rules of the symbol system). Their truth has nothing to do with the way the world is (we don't think of ideas as literally having colour, unless perhaps we have synaesthesia). They would be just as true in any world, as long as those symbols were used in the same way we use them.

Likewise, some of our beliefs do derive from our perceptions and are amenable to evidence-testing. It may be trivial to say that as I write, I believe myself to be looking at a computer monitor, but it is true—and when I turn my head my belief will change accordingly. Many of our beliefs are less directly related to our sensory inputs, but are nevertheless affected by them. If you believe that today is Friday, and the morning news depressingly assures you it's only Thursday, you will correct your belief accordingly. Pleasant though it may be to operate under the delusion that the weekend is not far off, knowing what day it is has advantages. Accurate beliefs allow more accurate predictions, and these in turn give us greater control over the world, as well as some protection against disappointments. Evidence-testing has led to huge advances in science and medicine. The truths it uncovers are contingent, in that they depend on the state of reality and on our ability to monitor it accurately.

If we rigorously tested every belief we encountered before accepting it we would all be perfect positivists, and the correlation between what we think is going on and what is actually going on would be much greater. We would also be terminally insecure and incredibly dull. So much of what we believe is taken on trust that imposing an evidence criterion— even the extended criterion of accepting any beliefs confirmed by science in general—would shrink our cognitive horizons to a tiny, unimaginative sphere. There's an awful lot scientists haven't got around to testing yet. All the most exciting beliefs in human history would be discarded as too difficult (or impossible) to test.

Excitement, of course, can be positive or negative, and some exciting beliefs are also dangerous. But the way to tackle that problem is not to insist, as some rationalists seem to, that human beliefs should all be based on evidence and reason. That is ethically dubious, since many 'irrational' beliefs do little or no harm and provide a lot of comfort. It is also unrealistic, not least because it tempts us to raise science, and scientists, onto the pedestal of authority—an unwise location for something which should depend on uncertainty, not dogma. Education and scientific training may help to open some minds, but they are not sufficient. There are numerous highly educated and/or science-trained bigots around, ready to dismiss alternative points of view without a second glance, never mind an examination of the evidence, because they feel so strongly that they are right. Until we attain the capability to redesign the human organism, unfounded misconceptions are here to stay.

One of the ways in which people strengthen their favourite unfounded misconceptions is to lower their standards for evidence (conversely, disliked ideas risk icy scrutiny worthy of the most sceptical philosopher). If Jim, who believes in ghosts, swears he heard strange noises one night in an old, empty house, he is basing the belief that ghosts made the sounds on very weak evidence and ignoring other explanations. Jim may not always treat such evidence as valid—in his modern flat he may disregard miscellaneous creakings and groanings as due to the heating, the plumbing, or the neighbours. But if he has chosen a world-view encompassing ghosts he needs support for this belief, so in places where he thinks ghosts are likely to lurk his interpretations change accordingly.

SCIENCE AND CERTAINTY

Scientists, on the whole, have not adopted Jim's world-view; but scientists cannot 'prove' that ghosts don't exist. Brain workings are not about generating certainties but about adjusting probabilities. Science, one of the most effective products of the neural ocean, 'proves' nothing; for proof means certainty, and certainty belongs to the realm of logic and mathematics. Science is a more human endeavour. Its aim is better theories, not perfect ones; explanations which are a closer fit to reality and so can generate more accurate predictions. Some explanations provide a very good fit, such as evolutionary theory; some, like astrology, offer a very poor fit to the way the world is.[1] However, what works for me may not suit you so well; and although arguments of accuracy, efficiency, simplicity, and reasonableness *ought* to carry great weight, they are often brought up short by the sheer power of human belief. If even signals from reality cannot always shift the brain's expectations, we should not perhaps be surprised that mere arguments fail.

Evidence and reason give us much, but they do not by themselves give us reasons to care greatly about the beliefs they support. (Love of truth is an often-cited motive—and often-exaggerated too; like most kinds of love, it is rarely as pure and unconditional as it seems.) For example, I believe that the synapse theory of how neurons talk to each other is extremely interesting, important, and influential. Yet if someone were to convince me that this theory was wrong my heart would remain unbroken. When I come across scientific articles suggesting that the theory may need amendment, I don't weep bitterly (nor do I fling the papers aside in disgust and set out to plant a bomb in the authors' laboratories).[2] Intellectually engaged I may be—years of my life have been spent on neuroscience, after all—but I have not invested huge emotional resources in the synapse theory.

This is the scientific stance at work, attempting to minimize emotional commitment to theories and hypotheses which may well be overturned. Emotion-laden beliefs do creep in, of course, especially when issues of status are involved; but scientists who get too attached to their theories risk derision from their peers. As noted earlier, failure and revision are acceptable parts of the 'ingroup' stories that scientists tell about themselves.

Whether we care about a belief is crucial to whether we act on it. Our concern depends on where the belief has come from, how strong it is, and how we feel about it. Let us consider each of those in turn.

Sources of belief

Patterns of neural activity, as we have seen, are triggered by and correlate with signals from three sources: other patterns, external sensory inputs, and inputs from the body. In principle, for any single neuron in the brain, we could measure the proportions of its input which come from the world directly, from signals internal to the body, and from other neurons. These can be thought of as competing for influence over the neurons to which they send signals. When signals from the world are strong and clear (like the sudden sound behind you which wrenches you out of that daydream you were enjoying), they will drive 'bottom-up' activity. When they are weak or ambiguous, however, the role of 'top-down' interpretation increases. And when they are accompanied by changes in body state, they can, as we saw in the previous chapter, become considerably stronger.

Strong beliefs mean more to us than weaker ones. This 'meaning' involves four interdependent aspects of the belief: its importance (how much of our thought and action is based on the assumption that it is true); its connections with other ideas (the semantic 'richness' of its associations); its closeness to sensory inputs (the abstractness or concreteness of the belief, its testability and dependence on reality); and its links with internal signals (its emotional meaning, the value it has to us). Beliefs which rely on sensory perception tend to be weaker, because the world which generates them can and does change. Beliefs which rely on reason are at risk from errors in reasoning; cultural change, for instance, can transform one generation's obvious assumptions into objects of derision for the next. Beliefs associated with strong emotions do not need to track changeable reality, nor do they need to be disturbed by errors of thinking. Because their power is fuelled by the third source of neural signals, the visceral body, they need only change when feelings change—which may never need to happen. Consequently, they can become extremely strong and highly inflexible.

Strength of belief

The strength of neural patterns of activity depends, as we saw in Chapter 5, on the strength of the underlying synapses (the probability with which signals transfer between neurons). The biochemical processes which allow synaptic strength to change are extremely complex, and describing them would take more space than I have available.[3] Suffice it to say that synapses can alter their chemical balance so as to make signal transmission more probable, and that this strengthening tends to happen when the neurons they connect are repeatedly co-activated ('neurons that fire together wire together', as the saying goes). This may occur when one neuron triggers the second, or when other signals trigger both at once. Conflicting patterns, or any form of inhibition, will interfere with co-activation, and hence with synaptic strength. Consistent patterns, frequently repeated, will enhance it.

The beliefs we like best are those which cause strong, clear patterns of neural activity which fit easily among the patterns we already have. They are 'like us', and we generally prefer what is like us, especially since accepted beliefs become components of our overall identity. They should add something new (or why bother?), and they should if possible untie or bypass an existing knot in our networks, as that will reduce the unpleasant sense of conflict and make us feel better. If they are to form stable patterns, they should be clear, consistent, and relatively simple. They should also come with an emotional booster to help carve their impression into cortex.

Otherization relies on beliefs which meet all these requirements. The core messages may be stated in many ways, elaborately and with apparent casualness, but at base they tell a clear and simple story with three core beliefs. The first is: these people are different, disgusting, not like you (not quite human). The second is: these people want to harm you, or have already harmed you or people like you. The third is: removing these people will solve your problems. Skilful otherization presents both doctrines and proponents with an eye to the intended audience, so that leaders can appear as 'one of us', 'a man of the people'. Thus their notions acquire that sheen of gentle reason we tend to associate with people like us. Otherizing stories also provide listeners with explanations of problems which they themselves can neither explain nor resolve—ideology as therapy—and this reduction of mental conflict is a great part of their appeal. Importantly, such stories are framed in familiar language, made to fit the local culture of their listeners; this eases the key ideas more gently into brains which might otherwise balk at their deadly implications.[4] And from the rousing brass and mass euphoria of Nuremberg to the ferocious language of Rwandan radio, otherization makes effective use of strong emotions.

How we feel about beliefs

The emotional meaning of a belief depends on what emotional input is present while the belief's neural pattern is activated by signals flowing through it. This input may relate to the content of the belief: for example, the claim that rats are disgusting may activate patterns involved in feeling disgusted (possibly all the way down to the brainstem and stomach, for those who really have a thing about rats). But the emotion may have no such link with the belief and yet still influence it. If you learn to your horror that rats have been seen on a street where immigrants live, you may in future (unconsciously or otherwise) associate your negative feelings for rats with immigrants—especially if you already dislike immigrants. It may be that your own, much nicer neighbourhood also has rats; it may even be that the rats have been attracted to the area not by the immigrants but by the food your neighbours put out for local birdlife. Yet the taint of disgust, once applied, is hard to dispel. (It is also easier to blame immigrants than to jeopardize relations with your neighbours.)

We often think of beliefs as neutral entities, denizens of the cool realm of reason. But beliefs can carry weighty emotional baggage. Apart from the effects of context mentioned above, there are two main sources of this added value. The first is that many of our beliefs originate from other people, people about whom we often have strong feelings. The second is that beliefs themselves have evaluative implications because they are predictions about the way the world should be.

Predictions which are confirmed, or contradicted, are not simply neutral, because humans aren't simply computers. Confirmation strengthens our sense of control and is rewarding; contradiction undermines us and makes us feel worse.[5] For beliefs we don't much care about, the effects are miniscule. For our dearest beliefs, however, they can be transformative.

Beliefs are expectations

Beliefs, like all mental furniture, are grounded in neural patterns of activity. To be articulated in thought or word as specific propositions, beliefs must activate the particular patterns which ground the associated symbols. Yet beliefs need not be made explicit in this way in order to affect behaviour. As we saw for 'shadow' emotions in the previous chapter, activation of a neural pattern need not include the activation of every participant in that pattern. It may be useful here to think of beliefs as expectations: predictions about the way the world is likely to be. Just before you reach for an object, you can be said to have all sorts of beliefs about how your arm will feel as you move it, what shape your hand should be to grasp the object, and so on. Your brain continually generates these predictions about your interactions with objects. Usually they stay unnoticed unless they turn out to be wrong. If prompted, you may articulate some of them in language; most of the time there is no need to do so. Symbol systems provide the common currency which allows you to explain yourself, either to yourself or other people. Lacking that communicative motive, most of our predictions do their work unrecognized.

Viewing beliefs as expectations allows us to expand the category of beliefs beyond the traditional philosophical examples of consciously expressed statements. There is now plenty of research suggesting that beliefs need not be conscious to affect behaviour.[6] If beliefs are predictions, they need not be statements either, although they may sometimes be expressed as statements. As expectations, beliefs are also revealed as bridges between the real and the potential, linking past and future worlds. Your belief that the sun will rise tomorrow relies on a life's-worth of days on which the sun has never failed to rise. Created by experience, your beliefs contain hypotheses about what may be as well as memories of what has been.[7]

More abstract and complicated beliefs can be thought of as clusters of associated predictions. For instance, the belief that the ancient Eygptians worshipped the sun yields the expectations that, should you study the relevant historical material, or hitch a ride with a time-traveller, you would find evidence of sun-worship in ancient Egypt. (These claims are in turn made up of lots of predictions about the presence of discs in temple and burial imagery, references to the sun in religious rituals, and so on). In principle, these predictions could be used to test the hypothesis, perhaps by studying research on Egyptian rituals or going to look at temples and pyramids. In practice, we do not have time to check everything ourselves. Many beliefs are therefore lifted, more or less wholesale and with more or less adjustment, from other people.

Social beliefs

We are all dependent on others for information about the world that we do not have time, energy, or expertise to check ourselves. More fundamentally, we need other people to teach us how to use language, how to name and understand our feelings and behaviours, how to manage social interactions, and how to distinguish acceptable from unacceptable beliefs, ideas, customs, and conventions. Thrown into uncertainty by a confusing and capricious world, we seek the views of people we trust, and use their verdict to strengthen our own opinions.

Viewing beliefs as predictions has another consequence. Philosophical work on beliefs has emphasized the issue of their truth as statements ('propositions'), but beliefs may not always be expressed in propositional form. Predictions may turn out to be true or false, but first they must be tested. Beliefs are working hypotheses, and we tend to work with them until told otherwise. Beliefs which are never tested, or indeed could never be tested, may nonetheless be widely accepted. In such cases the key question becomes not 'Is it true?' but 'Has it been challenged?'. Whether we trust in a belief, or in a person, may have remarkably little to do with that belief or person's relationship to truth, as long as our trust in them has never been questioned.

With respect to sensory perceptions, we tend to trust them more if they are consistent over time and space and if the perceptions match up across different senses. A solid, touchable person whose path through space is continuous and whose visual image is opaque and does not flicker is probably a real person, not a hallucination. The physical environment, our universe being a relatively placid place, does not present human beings with logically contradictory inputs (at least until they evolve to the extent of developing quantum mechanics, whereupon they will be faced with waves which are also particles, non-local communication, and all the other counter-intuitive wonders of the subatomic scene). It is therefore safe to assume that any conflicts between patterns of neural activity are, putting it crudely, our fault, not the universe's. The stress of these clashes, and the need for control which drives us to remove it, motivates our acquisition of ever more accurate and trustworthy theories of how the world functions. Consistent stimuli are trusted; inconsistent stimuli signal an error. Moreover, because conflict is unpleasant, we tend to associate unpleasant stimuli with conflict and pleasant stimuli with trustworthiness.

Social perceptions carry no such guarantee, because individuals can and do conflict. Yet we tend to rely on the same rule which emerged from our interactions with physical reality: trust consistent and pleasant inputs, be wary of changing or unpleasant ones. We therefore tend to trust news from multiple sources more than news from one, information which matches our expectations more than challenging information, and claims from people we like or respect more than claims from people we don't. How much we trust the views of one particular person on a topic depends on a number of factors, including how much we like them, what they mean to us, how reliable they have proved in the past, and what other people we trust think of them. It also depends on their social status and expertise. Someone who has demonstrated power and knowledge (for example, by gaining the title of 'Professor') is assumed to be better than we are at predicting and controlling the aspect of reality in question.

Unfortunately, our beliefs often rely heavily on the very same ingroup members to whom we turn when those beliefs are challenged, because they are typically the people from whom we learned our beliefs in the first place. As independent witnesses they can therefore leave a lot to be desired. Furthermore, group dynamics can produce considerable pressure on dissenting members, artificially inflating consistency.[8] This can leave members with a misleadingly positive judgement of the social 'weight' behind an idea. If everyone (at least everyone who matters) appears to agree about, say, a particular view of women or black people (because they have not openly disagreed), then where are the conflicting signals which might prompt someone to re-examine those beliefs or take on new ones?

Social support is particularly relevant when people ask whether beliefs and the actions arising from them are justified. This means more than simply having good reasons for them (however one defines 'good reasons'), because acting or believing without good reasons is not a neutral activity; it makes us uncomfortable. A person who acts on beliefs which they know to be unjustified is liable to experience conflict, especially if they are made aware of reasons not to act (like moral inhibitions). They may nevertheless have strong desires to act, and these will lead them to search for justifications.

How do actions and their associated beliefs come to be justified? By seeking out reasons. In brain terms, these are patterns of activity which re-route neural signals, draining them away from the point where they conflict, and hence reducing the uncomfortable feelings which conflict stimulates. For instance, talking about acting cruelly with people who react favourably can legitimize the idea of cruel action, especially if those people are liked and respected. Their approval is experienced as rewarding, which counteracts the negative conflict signal with a positive one. Otherization achieves the same effect. 'Hurting people is wrong, yet I want to hurt these people', points to an easy solution—they aren't really people.

Beliefs and identity

Our strongest beliefs matter immensely to us, and challenges to them can seriously damage our health. They dominate our cognitive landscapes and make a major contribution to our sense of who we are. We all have beliefs which are so important to us that if they were suddenly to be reversed we would think of the result not as 'me with a different belief' but as a different person. It is hard to imagine Hitler without his anti-Semitism. Because our strong beliefs are so large a part of us, we defend them against challenges as we defend ourselves: by avoiding or ignoring the provocation, marshalling reasons which undermine its impact, or sometimes lashing out at the attacker.[9]

We identify, and sometimes entirely define, human beings by characteristic beliefs as well as particular behaviours. Which beliefs those are depends in part on how distinctive or novel they are: a man who is the only Mormon in his street is more likely to be tagged as 'the Mormon' by his acquaintances than a man whose neighbourhood is rich in Mormons. Cultural changes are also relevant to which aspects of other people's identity we choose to emphasize. In recent years, factors including the burgeoning political importance of militant Islam, the collapse of communism, globalization, and social pressures against blatant racial prejudice have drawn attention to people's religious and other cultural beliefs. Consequently we now speak of the rise of 'identity politics', in which—at least in theory—groups are defined not by profession, social class, or racial background, but by the beliefs and cultural contexts of their members.

Breaking away from the older categories of race and class (in the sense of a group one is born into) might seem a liberation, a step away from the essentialist ideologies which condemned those of a certain ethnic type or social origin to impoverished life-chances. Social origin and ethnicity are aspects of themselves which people can do nothing to change, forcing them into deceit and disguise (or social rebellion) when these become criteria for excluding them from life's advantages. Histories can be reinvented, accents, mannerisms, and physical appearance altered in response to social pressures—but at considerable cost to the individual. In principle, fairer societies are those which judge by criteria people can control, at least to some extent: education, experience, financial status, and one's 'cultural' beliefs. Identity politics, if indeed it does group people by culture rather than, say, race, should therefore give rise to a less unjust social system, and that indeed has been the claim of many proponents of the identity-based approach.[10]

But are beliefs so flexible? Yes and no. Beliefs about which we care little cost us little to change, but the costs of changing our strongest convictions would be horrific: as traumatic a prospect as physical amputation, or even more so, since changing such a belief would feel like breaking apart the entire self. That applies to all of us, not just the (secular and religious) extremists who hit the headlines because their beliefs require violence. In attempting to understand why 'fanatical' believers act as they do, we should perhaps bear in mind that in challenging their ideas we are in effect demanding that they change much of who they are, which from their perspective can look like psychological suicide. This does not mean their notions may not be ridiculous, unreasonable, or downright dangerous; that is a separate issue. The point is that strong beliefs at least may be closer to being core aspects of the self than they are to being easily adjustable features, which makes the costs of changing them inordinately high. (That excludes the 'exit costs' of leaving the community of believers, which can also be appalling; for example, when religious beliefs hold apostasy to be punishable by death.[11]) What seems to us like a simple mental adjustment may strike believers as a deadly threat to their identity.

Here the importance of groups is manifest. When believers define themselves first and foremost by their membership of a symbolic community (a group defined by its members' shared and cherished beliefs), personal identity is built on the assumption of the group's continuing existence. Groups, however, are not people, because groups don't have bodies of their own. A symbolic community may gain a kind of body by developing physical institutions (like the White House in Washington or St Peter's Cathedral in Rome). Its identity then resides in part in the presence of those buildings and the communities which use them. Without this embodiment, however, a group's symbolic identity and its physical existence are one and the same. Both will continue only as long as there are faithful adherents. What matters is that the beliefs are preserved, not which bodies preserve them.

This dependence, for both groups and members, is also freedom. Groups without physical institutions have less to lose and can afford to make fewer compromises when they face threats. Direct-action pressure groups, guerrillas, paramilitaries, and terrorists are ostensibly less powerful than their enemies. Yet as portable weapons grow increasingly destructive and the media's usefulness for shifting public opinion becomes apparent, small groups are finding that they can wreak havoc—even when their opponents are well-armed states or powerful organizations.

As for group members, outsiders often see the believer as a fool surrendering precious autonomy, lessening his individuality by making a shared creed such a large part of his identity. But in return the group takes on many of the burdens which individuals might otherwise have to face alone. Anxieties related to mortality, in particular, can be greatly relieved by group membership.[12] Believers know that the group-related sections of their identity, which not only matter hugely to them but *are part of them*, will continue to exist after they themselves have died, as long as the group continues. Immortality, even partial and uncertain immortality, is a considerable incentive—and many groups, of course, explicitly offer beliefs about life after death in their ideologies. Challenging the ideas of such a community attacks both its identity and existence, as well as threatening the individual identities of group members and their hopes of a future (symbolic, if not physical) existence. When symbolic survival is more important than physical survival, as it can be, it is hardly surprising that believers sacrifice themselves to preserve the beliefs they cherish, knowing that someone else will keep the vision alive.

Threat and defence

Many of the ideas we encounter every day strike us as harmless, pleasant, or barely noticeable. Some are undoubtedly beneficial, bringing new ways of understanding the world, enjoyable emotions, or useful social credibility. Some trigger only the briefest of neural disturbances (a friend of mine has the enviable ability to read a magazine or watch commercial television without remembering any of the adverts). Some, however, conflict with the established contents of our cognitive landscapes, and as such can be considered threats to the mental status quo. They carry costs, not benefits: instead of fitting smoothly in with the patterns we have already, they demand that some of our mind must change to fit their truth. By doing so, they present an alternative to the way we do things, in effect announcing that at least some of the correlations we have learned to rely on are untrustworthy. The more important to us the beliefs being challenged, the more stressful the challenge. Changing well-established patterns is effortful, and the stronger the pattern the more effort it requires.

Meme theorists have depicted our cultural environment as a frenzied struggle for survival, with every notion, or 'meme', fighting to grab our attention. Those memes we do take on board, like viruses, stealthily conquer the brains they need to reproduce, prompting us to laugh at the joke and repeat it, use the neat soundbite, or brainwash our children with the false beliefs.[13] However, you do not need to go the distance with the virus analogy to think about the cultural transmission of ideas (which is as well, since memes and viruses have many differences when you look more closely). Anthropologists, for instance, have been researching how notions spread for years without the need for memes.[14]

Nonetheless, the metaphor of ideas as pathogens is a potent one. Ever since the development of the germ theory allowed us to understand how infections spread, propaganda has made the link explicit by comparing unwelcome ideas to dangerous diseases. The French police spy Lucien de la Hodde, writing shortly after his country's 1848 revolution, called his government's opponents 'the very leprosy of the body politic' and 'a virus with which France has become inoculated'.[15] Mao Zedong, whose choices led to millions of deaths, claimed that 'Fighting against wrong ideas is like being vaccinated'. Adolf Hitler, emphatic prophet of social hygiene, argued that German pre-war culture was weakened by decadent 'trash' ('pestilence, spiritual pestilence, worse than the Black Death of olden times . . . these scribblers who poison men's souls like germ-carriers of the worst sort'). Lest you should think the metaphor is only to be found in radical politics, we have more recently seen an Oxford science professor keeping up the grand tradition by calling religious faith 'a bacillus', a form of mental illness 'comparable to the smallpox virus, but harder to eradicate'.[16]

Note that not all of these quotations explicitly apply the metaphor of infectious disease to ideas *and* to the people who promulgate them. Yet all that tells us is that some propagandists aren't prepared to state the obvious implication of their arguments: that people who have 'wrong ideas' are plague-bearing organisms and ought to be eliminated for the sake of public health. Whether political, artistic, or religious, 'wrong' beliefs are often strong beliefs, and thus extremely difficult to remove by education, or, for those already infected, re-education.[17] It is much cheaper and easier to remove the people who hold them. Religious faith, for example, has survived the convulsive agonies of Democratic Kampuchea, the stifling years of the Soviet regime, and the supposed death-by-indifference of Western secularism, so any atheist serious about 'eradication' will need to use more extreme methods than a few adjustments to school curricula.[18] Fortunately for those infected with the faith bacillus, the atheist threat is not a serious one at present. The language of some 'brights' may be irresponsible, but their more thoughtful representatives advocate liberal ideals, including non-violence. So do many religious leaders, though that has not stopped religion from making its own extensive contributions to various mass murders around the globe.[19]

INFECTIOUS IDEAS

An idea, compared with a tiger or tsunami (or even a bullying boss), seems a poor, weak thing. How could so insubstantial an incarnation in breath, ink, or neural excitation put our lives at risk? We feel, rightly or wrongly, that we have a good deal of control over whether or not we 'buy into' new ideas. Threatening notions, of course, may not seem threatening when we first encounter them; only when we reflect may we realize their implications and the extent to which they conflict with our beliefs. Yet even then, an idea by itself cannot change our minds, let alone destroy us like a predator or natural disaster. Once we perceive the danger, surely all we have to do is close our minds to further contact.

Threats in the symbolic world of ideas, in other words, resemble pathogen or poison threats (disgust-threats) more than they resemble fear- or anger-threats. Both germs and notions are physically powerless, often hard to detect, relatively slow-acting, and only sometimes deadly. Unsurprisingly, the methods we have evolved to deal with dangerous diseases—learn the warning signs, avoid the source, quarantine the infected, and expel the contaminant—are the ones we use against dangerous ideas. Our emotions have also been co-opted to this new use, which is why beliefs we happen to dislike are so frequently described as disgusting. It is also why, throughout history, carriers of new and challenging beliefs—such as books and, far more importantly, people—have faced hostility born of revulsion, often with lethal consequences.[20]

Faced with a source of infection, like a corpse or excreta, we want to dispose of it (or, these days, let the appropriate specialists do so) with as little contact as possible, either by burying it, or burning it, or letting water wash it away. Ideas, of course, cannot be rinsed or burned away (though the concept of 'brainwashing' rapidly caught on); but the books and papers which carry them can be burned, and we may try to bury them.[21] But the people who introduce us to threatening ideas, or express their dangerous beliefs, can be disposed of via earth, fire, or water. Auschwitz, in the Nazi world-view the *anus mundi* where disgusting social contaminants like Jews and gypsies were excreted from the world, was a modern expression of a very ancient instinct.[22]

The conflict stirred up by new ideas can be so devastating only because so much of human life is lived in the symbolic world. What we refer to, when we use the terms 'I', 'me', or 'myself', is a human person defined not only by a body's physical boundaries but by symbolic aspects too: commitment to certain beliefs, particular ways of acting and thinking and using language, a preference for some ideas and dislike for others. As symbolic selves, each of us is a hugely complicated knot of beliefs, desires, and other neural patterns. Many strands of the tangle can also be found in many other people, but each is rendered uniquely ours by its place in our cognitive landscape, the mesh of connections which holds the knot of self together.

Beliefs which challenge us by being different from our expectations tend to be experienced as particularly stressful if we are already under stress for other reasons. Even academics, those insulated creatures, feel the difference. Leisurely contemplation of a new theory is one thing; having to defend your own ideas in a debate is quite another. Nuances tend to vanish in the heat of argument, as emotions boost and harden our neural patterns and we seek to defend what matters to us. The more important the beliefs we are trying to protect, the more we will dislike the attacker—unless we are secure enough not to feel threatened in the first place. If, however, we feel vulnerable for other reasons, we will be all the more motivated to maintain our own beliefs and reject conflicting ideas.

Moreover, the kinds of abstract belief we tend to cherish often need defending in a way that beliefs closely tied to the world do not. For the latter, reality can serve as an independent arbiter to decide the question of whether, for example, it is currently raining where you are. Abstractions, however, depend on other abstractions and on the person's past and present feelings; for many of them, reality has no vote. Beliefs to which we become passionately committed can escape any ties to the world they may previously have had, allowing the believer to ignore or reinterpret any contrary evidence in his or her favour. Unfortunately, though this data-massaging may suffice to reassure believers, it may not convince a sceptical challenger.

This picture of a pair of opposing minds, each with its mesh of patterns, each trying to out-argue the other by hurling justifications across the void between them, is somewhat misleading. Two people may find themselves isolated in confrontation on one specific occasion, but they bring to that encounter a wealth of social experience—and support. We are more likely to believe a new claim if the person next to us believes it too, just as we tend to like people more if they believe the same things we do. If something new happens, we tend to turn to other people to see what they think of it and whether they are reacting as we are, shaping our perceptions and treating the event as more or less real accordingly.

World-shaping

I have referred repeatedly in this chapter to people changing their beliefs in response to incoming signals from the world—including the symbol-ridden social world provided by other people. In earlier chapters I emphasized the importance of seeing how the brain works in terms of action-selection, rather than viewing it as a nifty thinking machine. Our increasing ability to react effectively and in complex ways to complicated situations, and to predict and control those situations, has given us the leisure we need for fantasies and daydreams, and many other forms of creativity besides; but the pressures which crafted human evolution had more to do with facilitating the actions which helped our species to survive. That imperative to compete and reproduce is still with us. Twisted into a myriad cultural forms, and fuelled by the poisonous power of social pressures, it drives our pursuit of many impossible ideals: perfect beauty, perfect health, perfect relationships producing perfect children, and so on.

World-shaping is what you get when you put a person's habitual urge to act together with their commitment to particular beliefs. The stronger that commitment, the less inclined the person is to change their belief, even when incoming signals clearly contradict it. There are strategies one can adopt to ignore, downplay, or reinterpret those signals in order to make them less threatening and reduce the conflict they cause, but this can take nearly as much effort as changing the beliefs, particularly when the challenging signals are persistent. And effort is a serious disincentive. If brains are thought of as action machines, not thinking ones, then energy conservation demands that they minimize effort by not thinking unless the situation demands it. When unwelcome ideas require extensive and effortful thought to accommodate them, it becomes easier not to bother changing the brain, but instead to shape the world to fit— by removing the irritant source of the challenging signals.

BELIEF CONFLICT

How do brains process conflicting signals? Neurons assess the positive (excitatory) and negative (inhibitory) signals reaching them at a point in time and react (fire off their own signal) if the balance of their inputs is sufficiently positive to surmount the cell's elecrical threshold. Incompatible signals, marking events which cannot go together in the world, or incompatible beliefs (for example, 'causing pain is wrong' together with 'slapping the kids is OK'), tend to excite patterns of neural activity linked by inhibitory synapses. Activating one of these (which may lead to it becoming explicitly conscious) will tend to suppress activation in the other, and vice versa. The greater the inhibition between the two neuronal networks involved, the more incompatible the beliefs concerned.[23]

Incompatible beliefs are like territorial animals. Kept far enough apart in the cognitive landscape, they need never interact. Force them together, however, and conflict ensues. The weaker will change to accommodate the stronger, with the degree of change reflecting the disparity between them. A new belief, like someone moving into a new area, must run the gauntlet of neighbourly assessment: its 'fit' with a person's established ideas. The less it conflicts—the more compatible it is—with those ideas, the more easily it can merge into the cognitive landscape.

For example, if Jack fancies Jill and believes that any woman he fancies must fancy him, and Jill thinks she doesn't fancy him (but might find her standards slipping if she had enough to drink), they cannot both be correct, in our real world. Yet the degree of incompatibility is minor, and with alcohol, excitement, and the pressure to pull, Jill's weak beliefs about Jack's desirability can easily be altered. If, however, Jill is firmly of the opinion that she'd rather date a cockroach, the environmental pressures may reduce her resistance to Jack's advances, but they are less likely to change her belief in his repulsiveness. What options, other than rape, does this leave Jack?

If Jack's two beliefs—that he fancies Jill, and that all the women he fancies fancy him—are both very strong, he may persist in trying to change Jill's opinion of him. (As noted in Chapter 4, inputs which don't fit our expectations can get short shrift from early on in brain processing, giving us remarkable capacities for self-delusion.) If not, he is likely to amend or discard one or both of them. He may, for example, modify 'all the women' to 'most', 'some', or, more realistically in most cases, 'a few'. He may decide that Jill is not a 'proper' woman—that she is frigid or a lesbian, for instance. Or he may decide he doesn't fancy her after all, perhaps resorting to verbal abuse to reinforce this newly acquired belief. Alternatively, Jack may keep both original beliefs intact by using denial: ignoring the evidence that Jill doesn't fancy him (that rude remark wasn't intended, she was drunk or showing off), and perhaps avoiding her thereafter, to minimize the threat she poses to his ego. Finally there is the risk of world-shaping by violence: either forcing Jill to have sex or assaulting her to make her less attractive. Which option Jack takes will depend on the relative strength of the beliefs involved. Changing a challenged belief always involves a trade-off between the stress of conflict and the effort of suppressing old neural patterns in favour of new ones. The stronger a belief, the greater the motivation to keep it as is and alter other beliefs—or the world—instead.

THE HABIT OF CONTROL

World-shaping is something humans do incessantly. If something we see on the television 'stresses us out', we can and do switch off; if a tree is in our way, we may choose to remove it, no matter how beautiful or healthy it may be; if an insect invades our home, or a dangerous animal our neighbourhood, we casually destroy it. Humans have drilled through mountains, created gardens from deserts and vice versa, drained lakes and flooded entire landscapes to suit their requirements. You may not personally have done any of these things, but you engage in small-scale world-shaping too: every time you move an object from where it is to where you think it should be.

Some objects and events, of course, are beyond your control. No matter how hard you hope for sunshine on Saturday, you have no influence over whether the sun will shine. The same is true of the social world, in which many of the most important ideas and their bearers are beyond your reach, despite all the interactivity and democracy claimed for our newly webbed world. Social symbols and their carriers, however, tend not to have the stubbornly undeniable independence of, say, a rock, tree, or earthquake. Social authorities can be undermined, for example, by implying a venal hidden agenda, or presenting a conflicting authority.

Social world-shaping, like its physical counterpart, aims to negate the sources of conflicting beliefs. Because it takes effort and risks retaliation, it tends to be directed against less powerful individuals and groups. The bigger the power disparity, the easier the source is to target with some form of hostile action. What that action will be depends on the beliefs of the world-shapers. As with any action, they will rely on predictions of the likely consequences, aiming to resolve the problem effectively while wasting as little energy as possible. However, estimates of how effective an action will be are harder to get right than estimates of how much energy that action will take. So world-shapers intent on using otherization to remove those with challenging beliefs often begin with less drastic actions, such as social distancing and ostracism, hoping to save energy and achieve their goals at lower risk to themselves. Only after these have not had the desired effect (removal of the social irritant) will more extreme measures be taken: active hostility, bullying, harassment, and so on up to, at the negative limit, elimination or destruction by any means available.

In seeking to world-shape, a person's expectations about how reality should be will affect the way in which they try to change it. An obstreperous gadget failing to work as expected does not as a rule provoke fear, grief, or a sudden cheerful grin; it provokes the urge to hurl it across the room. If we succumb to that urge, we symbolically expel it from our presence. We also, quite probably, break the thing, which inconveniences us, but does on the other hand bring the object back into line with our expectations. Broken gadgets can be expected not to work; unbroken gadgets should work; gadgets which should work and don't are infuriating misfits, to be treated accordingly.

As for gadgets, so, it seems, for people. Many of our expectations are mild and can readily be changed, but some are so much part of us that they provoke us to act in the ways most likely to change the world to fit them. People who ought to be subordinate but do not act appropriately are made to feel their inferiority (and we are exceptionally good at that). People whom we thought worthy of respect who then disgrace themselves are mercilessly pulled down from their pedestals; we preserve our ideals by disengaging them from the sinner and finding a substitute as fast as possible. Women, who ought to be interested in marriage and babies, face prejudice when they pursue careers instead; the more strongly traditional their society's view of their role, the rockier their paths are likely to be. Finally, and most relevantly to the topic of this book, people labelled as weak, treacherous, selfish, or disgusting 'others' are most threatening when they challenge those stereotypes—by displaying strength, being trustworthy, showing kindness, or evoking the instinctive warmth and concern required, as anyone who has changed a nappy knows, in order to overcome disgust.[24]

EXPLAINING THE INEXPLICABLE

In his book on the Holocaust, Lawrence Langer describes the testimony of a survivor of the Jewish ghetto of Kovno in Lithuania. During a *KinderAktion*—a round-up of young people for execution—the survivor testified to being 'present in the room when an SS man entered and demanded from a mother the one-year-old infant she was holding in her arms. She refused to surrender it, so he seized the baby by its ankles and tore the body in two before the mother's eyes'.[25] As examples of gratuitous cruelty go, this is among the most extreme, capable of making us feel physically sick, or moved to tears, or both.

World-shaping may help us make sense of what at first sight appears incomprehensible. Why did the SS man kill the baby in *that particular* egregiously horrible manner? He could, after all, have shot it, or broken its neck, or thrown it against the wall. Many other infants have been dispatched by these methods, which take less time and energy than the one the SS man chose. However, in terms of world-shaping they are less effective, as they are less able to align the world to the SS man's expectations.

Consider the likely background of this man, about whom we are told so little. He must, we can assume, have been thoroughly familiar with Nazi ideology. The SS, we know from other examples, were taught to see Jews as repulsive, subhuman, dirty, and disgusting creatures. A baby is a desperate threat to that dogma. Its helplessness calls us to empathy, to protective and tender feelings. Even a screaming infant, nerve-jarring and frustrating, is still so obviously a human child, not the bestial, repellent thing which Jewish infants should be according to the Nazis. A baby dead from a shot or a broken neck, even a baby slammed against a wall, is still recognizably a baby, a threat demanding fearsomely difficult changes in the SS man's beliefs. But an infant torn in half becomes disgusting: a dismembered corpse, with all the foul associations such carrion brings with it. The problem, from the SS man's point of view, has been resolved. The Jew is now as Jews are supposed to be, and he can feel as Nazis ought to feel about their victims.

CLEANING UP POLLUTION

Examples of this kind of thinking abound in mass killings. One of the most notorious instances, as noted in Chapter 2, is the use of metaphors of health and disease, infection and contamination, to justify the Holocaust. The concept of genocide as healing, a last-ditch cure for a desperately sick state, may seem stomach-churning, but it is not unique to the Nazis. Similar language was used by perpetrators of the Rwandan genocide, by the Khmer Rouge in Democratic Kampuchea, by the Maoist regime in China, and so on. Often the metaphors include references to poor hygiene and the necessity to 'clean up' problematic elements—as in the Spanish Inquisition's insistence on *limpieza del sangre* (purity of blood), or the massacres in El Salvador described as *La Limpieza* (the Clean-up).[26] In the wake of Lister (antisepsis), Snow (the discovery that cholera was waterborne), Pasteur (pasteurization), and Bazalgette (creator of the London sewer system), the medical profession found that emphasizing hygiene could bring remarkable advances in public health. The link between curing and cleaning was thus an obvious one for ideologues to make, and had the further advantage of allowing them to tap into deep-rooted fears about dirt and pollution.[27]

References to disgust and contamination are characteristic of all extreme otherization. Yet the best-known and most extensively researched examples of medicine's lethal co-option to the service of genocide undoubtedly come from the years leading up to the Holocaust. Hitler was a master of the language of filth and cleanliness, and especially adept at merging it with well-established concepts of the nation as a higher-level organism (the 'body' politic). In the mid-1920s, long before he was able to put his racial theories into practice, Hitler described his era as 'inwardly sick and rotten' and suffering from 'a disease', the 'syphilisation of our people', presenting himself as the physician-king who had the courage both to diagnose the problem and prescribe the remedy.[28]

That remedy was two-pronged. On the one hand, the Aryan people had to be cured of their racial disease. To achieve this, their blood (the presumed channel of ethnicity, pre-Crick and Watson) was to be purified of alien components. Hitler did not disdain education: 'No boy and no girl must leave school without having been led to an ultimate realisation of the necessity and essence of blood purity.' But he recognized the need to use more radical techniques as well—and to apply them not just to 'foreign bodies' like Jews and gypsies but to members of the master race itself. An example is the T4 euthanasia programme, practised on mentally and physically disabled Germans until 1941, when public outcry forced its closure.[29]

On the other hand, the disease itself—the *lebensunwertes Leben* (life unworthy of life)—had to be attacked and eradicated. As Robert Lifton makes clear in his seminal work on the subject, *The Nazi Doctors*, medically trained individuals drawn into the Nazis' project were understandably prone to adopt clinical metaphors.[30] Thus Fritz Klein, challenged to reconcile his participation in the Holocaust with his role as a doctor, appeared to have no difficulty in accommodating healing and mass killing: 'Of course I am a doctor and I want to preserve life. And out of respect for human life, I would remove a gangrenous appendix from the diseased body. The Jew is the gangrenous appendix in the body of mankind.'[31]

Non-medics were also susceptible to the comforting concept of the social body, in which individuals, or entire minorities, could be seen as healthy or malfunctioning cells. Hans Frank, who trained in law and became the Nazis' governor of Poland, saw the Jews as 'a lower species of life, a kind of vermin, which upon contact infected the German people with deadly diseases'. Eliminating them, he said, would ensure that 'a sick Europe would become healthy again'.[32] SS-Hauptsturmführer Hans Bothmann, the commandant at the Kulmhof (Chelmno) extermination camp, similarly explained to new arrivals from Police Guard Battalion XXI that 'in this camp the plague boils of humanity, the Jews, were exterminated'.[33]

Primum non nocere—an injunction attributed to Galen and Hippocrates—expresses the medical ideal: 'first, do no harm.'[34] The gap between this sentiment and what happened to people deemed 'diseased' in Nazi Germany provides a particularly potent illustration of humankind's astonishing capacity for otherization.

Summary and conclusions

Disease metaphors are powerful drivers of otherization. They offer clear, simple and highly emotive explanations of a person's, group's, or nation's maladies, co-opting evolved threat responses to increase the motivation to act. They fit smoothly among accepted beliefs about disease, including the ancient fear of strangers as disease-carriers. They also depend on the idea of the group as a body, fostering positive emotions among group members and imposing pressures for loyalty and unity. This makes the group appear more cohesive and therefore more powerful, both to its members and to outsiders.

Given the choice between changing beliefs (or other neural patterns) to fit the world and changing the world to fit the beliefs, a person will take whichever option seems easier. The more important the beliefs in question, the more strongly they will be defended, the more extreme the emotions which a challenge will provoke, and the more violent the response is likely to be. When the belief involves disgust, and that perception is challenged, human beings—people like us—can respond with lethal and hideous savagery.

Some cruelty involves misjudged responses to social threats which stimulate anger or fear. Some, however, involves deliberate disregard of the victim's human status, because otherization has made that status almost entirely meaningless for the perpetrator. Callously cruel perpetrators motivated by strong beliefs may see their behaviour as unfortunate but necessary self-defence, protecting them against ideas and people which threaten vital aspects of their identity.

Not all callous cruelty is provoked by belief defence, however, and sometimes beliefs can be used instrumentally to justify aggressive behaviour—or even strengthened specially for the occasion. Greed, fear, revenge and its cousin resentment, competition over resources: these and other motives can both fuel belief-driven cruelty and be fuelled by its ideological justifications. Beliefs may offer channels for the energies of emotions, linking desires to violent actions. Yet circumstances can change even strong emotions to such an extent that formerly cherished beliefs can lose their force and come to seem delusional, like bad dreams, to those who held them. And there is also sadism, which seems to be more about the pursuit of pleasure than the world-shaping urge to preserve vulnerable ideas. Beliefs have much to do with the worst extremes of cruelty, but they need not be fanatically strong to do their damage.

These last three chapters have looked at how human brains mediate action, emotion, and belief. In Chapters 7 and 8 we will apply that understanding to the problems of callous and sadistic cruelty.

Chapter 7

Why are we callous?

Hard towards himself, he must be hard towards others also.
All the tender and effeminate emotions of kinship, friendship,
love, gratitude, and even honor must be stifled in him by a
cold and single-minded passion for the revolutionary cause.

(Sergey Nechaev, *Catechism of the Revolutionist*, 1869)

C ruelty covers a multitude of sins. My aim in this book is not to explain every type of cruelty in detail (that would be impossible), but to provide a theoretical framework for thinking about cruelty in general. So far we have looked at how to define cruelty, distinguishing callousness from sadism on the basis of the perpetrator's primary aim in acting. We have also considered the underlying mechanisms of three crucial components of cruel behaviour: deciding to act, experiencing emotions, and having beliefs. Now it is time to put these components together: to assemble our framework and see what it can tell us about cruelty.

The topic of this chapter is callousness. By far the more common form of cruel behaviour, it ranges from physically harmless but psychologically painful verbal viciousness, ostracism, and bullying to the physically destructive 'cold-blooded' instigation of ethnic cleansing, starvation, and mass-murder. It is these large-scale atrocities which tend to attract most attention. Thankfully they are rare; yet callous cruelty can do a mind-numbing amount of harm without killing anyone. Human beings can be gentle and altruistic, but we are also very frequently callous.

What is callous cruelty?

To recall the working definition outlined earlier: cruelty is unjusti-fied voluntary behaviour which causes foreseeable suffering (unwanted, aversive psychological or physical harm) to an undeserving victim or victims. It involves a voluntary decision to act in order to achieve a particular goal, or goals. For example, a soldier who complies with his superior officer's order to execute a prisoner may have the primary goal of displaying his efficient obedience—as well as wanting to smash the stubborn bastard, take revenge for dead or injured comrades, end a tiring interrogation, get out of this stinking, airless environment, and so on. As the soldier aims and fires he additionally creates many short-term, subsidiary goals concerned with moving the relevant muscles in the right order, most of which never reach consciousness.

With respect to cruelty, it is not always easy for perpetrators, victims, or third parties to tell how perpetrators' goals are ranked. Motives, as we have seen, are often unclear, and stopping to think is the exception rather than the rule. For an act to be cruel, however, one of its goals must have involved deliberately behaving in such a way as to cause foreseeable suffering.[1] The higher that goal of harm-doing is ranked by the perpetrator, the more sadistic the cruelty. For example, a guard who forced civilians into an overcrowded train bound for Auschwitz could be judged guilty of various degrees of cruel behaviour, depending on his psychological state:

Case 1: the guard knows nothing about Auschwitz; he has been told the civilians are being resettled in the East. He is polite but firm, neither taunting nor beating them, but showing no compassion either. Nevertheless, he has chosen to participate in forcing them onto the train, and he may reasonably be expected to know or guess that once on the train, their futures are likely to hold considerable suffering.

Case 2: the guard is aware of the passengers' final destination and their likely fate. He asks them to hand over valuables, saying these will not be needed, but does not threaten or abuse people who refuse him. He does tell reluctant passengers that they will be shot unless they hurry up.

Case 3: the guard taunts the civilians about their likely fate and beats those who are slow or who refuse to hand over valuables.

In other words, cruel behaviour ranges across a continuum which includes both callousness and sadism. Thoughtless, indifferent, and self-ish cruelty cluster near the callous end of the continuum. Towards the more sadistic end lies cruelty used to terrorize, in which the suffering inflicted, although not the terrorists' ultimate goal, is foreseen and desired as a message to their audience. Furthermore, cruelty may draw on mixed emotions, as when killers plead regrettable necessity—'I hate to do this, but it's us or them'—even when the 'them' in question is a group of unarmed children. These are perpetrators who know they are causing suffering, but for whatever reason downplay its importance in favour of other goals and tell themselves they are compelled to act.

Why is callousness part of the human experience?

The theory of Darwinian natural selection proposes that human beings, who have evolved to behave in ways which facilitate the transmission of their genes, will reap genetic 'fitness' benefits by living in kinship-based groups—if group members restrict harm-doing within the group, encourage members to act in sympathy with joint goals, and punish the free-riders who exercise their rights without fulfilling their responsibilities. Group members may risk death or expulsion if they continue to act contrary to shared goals (e.g. by free-riding), but they can also benefit from doing so. The resulting selection pressures favour individuals who can understand what others believe, desire, and intend to do, who can ostentatiously behave well in the presence of more powerful ingroup members, who can seize opportunities for selfishness when they present themselves, and who can conceal conflicting beliefs and desires from the group.

What does this tell us, other than that evolution, if personified, would probably be very proud of Niccolò Machiavelli?[2] One implication is that cheat-detection will also be selected. Cheats are parasitic on other people's resources, so being able to spot a cheat—and warn one's allies—is beneficial. Detecting selfish exploitation is easier, however, if there are clear expectations of how people should behave: moral rules. Deviations from socially acceptable behaviour have costs for group members (for example, the cost of punishing the offender and repairing any harm done, if that is possible). Teaching new members what they should expect, punishing rule-breakers, and trying to ensure that everyone has similar expectations therefore become priorities for harmonious groups. This leads to the development of group-specific moral codes, as well as a pressure towards having similar beliefs in general.

Moral codes evolved to protect kin, not strangers. When limited high-quality environmental resources forced groups into competition, competitor groups (outgroups) became a threat. When the danger was 'clear and present', threat responses would be triggered, leading either to one group's withdrawal or to intergroup conflict. Broadly speaking, that conflict could be of two kinds, depending on whether the groups were co-dependent in some way (for example, trading partners or related by marriage). If they were, it benefited both parties to settle disputes over power while minimizing the damage to themselves, leading to the development of ritualized (that is, rule-governed) battles.[3] Some of a group's moral rules, including conflict-resolution procedures, could thus be extended to other groups, provided the two partners shared enough mutual understanding to make each other's behaviour comprehensible.

Confronted with strangers, on the other hand, a peaceable ingroup might gain useful new trading partners—or risk great harm. Game theory suggests that, at least for simple computer models of such interactions, it makes sense to be tentatively peaceable to begin with, but to react aggressively if the other party shows aggression ('tit-for-tat').[4] If the strangers are weak, however, the ingroup could benefit from attacking them and claiming their resources, whether or not they are hostile. On our definition, as impartial observers, this would be callous cruelty.

If the strangers seem stronger, and aggressive, the ingroup could justify reacting with a moderate level of violence (that is, only as much as
was needed to deter the outgroup and preserve the ingroup). But this is
risky; the outgroup may come back with greater force in future. Violence
can also be hard to calibrate, especially in the heat of a fight—so the
ingroup's reasonable deterrence may be judged outrageously vicious by
those who experience it, making retaliation more likely. In which case,
the ingroup could save itself the costs of defence against future threats
by using excessive violence: either to make itself seem so dangerous
that no one dare attack it, or to remove the attacker once and for
all.

Applying evolutionary perspectives to human cruelty can shed light
on its puzzling aspects, such as the well-known tendency for callousness
to be easier at a distance. People trained to believe in democracy should,
one might think, find the murder of a victim equally disturbing whether
it is done bloodily in front of them or not, but in practice we are face-
to-face creatures and direct observation makes a massive difference.
Our ancestors evolved local contacts and perceptions, like the ability
to read and respond to distress signals, long before they were able to
think in symbols. Distance, whether supplied by ideology, technology,
or both, reduces the power of a victim's distress to inhibit cruel behaviour. The huge collectives, impersonal structures, and specialist divisions of responsibility developed by modern societies also reduce the
sense of imminent, highly negative social feedback which helped to keep
our nastier forebears in line. For third parties, victims at a distance—
especially when imagined in large, highly homogenous groups—are far
less real than individually weeping human beings. Like numbers, they
are represented as abstract entities, evoking only slight shadows of the
emotions we would feel if we actually saw their blood, grief, and fear.
For perpetrators, victims at a distance are easier to otherize, and treating
them callously causes less discomfort.

Why are humans so much more cruel than other species?

Many species have evolved responses to deal with common threats (fear
responses when the threat is irresistible, anger responses for powerful but
resistible threats, and disgust responses to deal with material containing

pathogens and poisons). Some species, including humans, have also developed the ability to control and direct these responses—up to a point. Emotion regulation is particularly useful for social interactions, both because it reduces physical assault and because it allows individuals to feign emotions they may not feel (or may not feel as intensely as they pretend).

Group living also placed an evolutionary premium on individuals who were better at predicting their social and physical environments, including the behaviour of others. In humans, the development of symbolic thinking allowed the capacity for prediction to reach far beyond the prowess of other species. Blossoming vistas of culture meant that ideas became important in themselves. One consequence of this extraordinary transition was to expand the number of ways in which human beings could be made to suffer, by expanding the range of entities we value. Now we live and breathe as symbol-eaters, not just physical organisms. We can love ideals, commit to beliefs, cherish dreams, and feel outraged by threats to symbolic constructs like flags or books. This means we have much more to lose than our less imaginative ancestors. A prisoner stripped naked may suffer physically (e.g. from cold) and financially (from damage to valuable material), but the greatest damage is psychological: the humiliating and dehumanizing removal of a large part of a human being's public identity. Swimming in an ocean of notions lays us open to many more threats and chances, pleasures and harms than our pre-symbolic forebears.[5] Other species also engage in world-shaping, but they make fewer (and, as far as we know, less abstract) predictions. They have far fewer reasons to be cruel, while we have more imagination with which to dream up sophisticated tortures.[6]

Threats arise when beliefs and reality clash, generating a desire (the 'need for control') to resolve the conflict. When the beliefs are very strong, or when reality is easily adjusted, it may be less effort to modify reality. Our strongest beliefs are so much part of us that threats to them trigger highly evolved threat responses—which do not always sit well with modern morals. When this world-shaping causes suffering, third parties who do not share the beliefs may fail to accept the believer's claim that his actions were justified, and hence describe him as cruel.

Why are we often callous to close kin and people we care about?

The basic argument from evolution suggests that human cruelty should predominantly be directed at outsiders, while altruistic behaviour is targeted towards (k)ingroup members. Yet altruism to strangers, even when there is no chance of the favour being reciprocated, is well known. So is altruistic punishment, whereby people use their own resources to inflict costs on those who break social rules. Should we therefore reject the Darwinian view? Not yet. As noted in Chapter 3, research suggests that altruism to strangers and altruistic punishment may both have benefits for the person's chances of transmitting his or her genes.

Cruelty to ingroup members, however, is also widespread—and being callous to friends and family seems at best counter-productive, at worst genetic suicide. How is such cruelty to be explained?

One factor is that, as the paternity-testing industry shows, the natural human capacity for kinship detection is far from automatic. Natural selection doled out no yellow stars or name-badges—and while behavioural cues (like shared mannerisms, or the length of time spent living together as children) and physical cues (like ethnicity and physical resemblance) are undoubtedly important, they do not always suffice to determine who is related and who the cuckoo in the nest.[7] Humans have developed sensitive cheat-detection systems to assess their partners' fidelity, but as Shakespeare's Othello learned, these are not always accurate.[8]

As well as behavioural and physical cues, judgements of one person's similarity to another also rely on symbolic cues—like having similar beliefs or shared interests, or simply being told about kinship links—because so much of human identity exists in symbolic form. Estimates of similarity advise us as to moral attitude: is this person kin, to be valued, or non-kin, a potential threat or competitor? Psychological similarity, like its physical, behavioural, and historical counterparts, can lead us to regard a person as, if not our long lost brother, at least crucially like us: symbolic kin.[9] We may *assume* that such people are well endowed with genes we share, but we *know* they share our values and ideas; in the arena of cultural evolution they rank as siblings—that is, as cooperative (until we find them competing with us, as siblings can). That can be enough to trigger empathy, and all the human decency associated with it.[10]

Similarity can be a powerful determinant of behaviour, drawing people of different ages, ethnicities, classes, and cultures cooperatively together. The sciences and religions, which have their own 'thick' cultures—that is, a complex set of defining values, behaviours, and attitudes which allow scientists (or religious people) easily to identify others of their kind—provide clear examples of the force of similarity.[11] A British scientist (typically male and middle-class, with university experience and lengthy professional training) may well feel at ease with his German, Chinese, or Pakistani counterparts, whether in the lab or the pub. He is likely to feel less comfortable with other, non-scientific Britons, especially if they are working-class or upper-class, because the shared 'British' culture is thin compared with the segregating force-fields of class and profession.

Many cases of callousness to kin fit the evolutionary logic expressed by William Hamilton.[12] Exploiting parents who can no longer reproduce, or killing children you think will not survive, allows limited resources to be concentrated on the vital business of genes, and both humans and other social species have been known to abandon or destroy very old or very young members of their groups.[13] Other cases of cruelty may result from the problem, noted above, of calibrating violence intended as punishment (especially if the perpetrator is intoxicated at the time, as often happens in domestic violence).

Callousness may also arise, however, from mistakenly negative judgements of similarity. When people find themselves in conflict, their sense of similarity necessarily diminishes. Differences between them, formerly ignored, become more noticeable.[14] This increased otherization adds layers of threat to the basic dispute, making it seem more salient to both parties and laying the groundwork for mutual loathing to escalate. A friend who starts to seem threatening may, in cognitive terms, still be labelled 'friend', but the emotional landscape delivers a different judgement.

Finally, as noted earlier, natural selection is rarely if ever a dictator, for humans at least. We have evolved highly aggressive threat responses to meet the universe's common challenges, but we have also developed an extra universe of potential reasons to engage in world-shaping, with and without violence. We can make mistakes, reacting more aggressively than circumstances require. It is no coincidence that many perpetrators of child cruelty see their behaviour in terms of punishment (that is, as justified discipline). To them, the child is not an innocent victim but a challenger, a destroyer of expectations. Close kin can threaten a person's beliefs and self-image, can disrupt an orderly world with demands which seem unreasonable, as much as or more than strangers can. Even when our evolutionary *and* moral logics demand restraint, we humans have evolved the ability, and the motives, to ignore, deny, and subvert both rationales.

What is the relationship between cruelty and empathy?

> Monster of men, oh what hast thou here done
> Unto an overpressed innocent,
> Labouring against so many, he but one,
> And one poor soul with care, with sorrow spent?
> O could thy eyes endure to look upon
> Thy hands disgrace, or didst thou then relent?
> But what thou didst I will not here divine
> Nor strain my thoughts to enter into thine.

> (Samuel Daniel, *The Civil Wars*)

Threat responses and their associated emotions evolved in response to common hazards. But they are costly, and when chronically activated they can damage the body. Inappropriate or unnecessarily long-lasting threat responses waste energy and risk physiological harm; they may also antagonize or even damage valuable others, especially in highly social species. It makes sense, therefore, to use the response sparingly and to be able to suppress it quickly once the threat has receded or has been recognized as a false alarm. Unsurprisingly, many animals employ, and are adept at detecting, a range of 'stop' signals indicating submission and/or distress (to neutralize anger-threat responses), or non-aggression (to neutralize fear-threat responses). In humans, these include both verbal communication and non-verbal signals such as gaze aversion.

Disgust-threat responses, which protect physical and psychological integrity by distancing the self from the disgusting object, subside as the distance between them is increased, just as fear subsides once the predator has gone. But like fear and anger, disgust also has a highly evolved social role: it shapes responses to evidence of disease in other people (skin damage, suppurating wounds, running noses, and so on). If the social role of disgust is grounded in the more basic threat response, one might expect that typical reactions to sick people would mirror those triggered by other disgust stimuli: avoidance, expulsion, or elimination. And indeed, this often happens when the sick people in question are outgroup members. (Thresholds for triggering disgust are lower for strangers and distant kin than for close relatives or the self, reflecting the increased likelihood that pathogens carried by strangers will be dangerous; if you live with your family you get used to their germs.[15])

Valuable members of society, however, are another matter. Getting rid of sick group members is hugely wasteful, given that many infectious diseases have thoroughly repulsive symptoms without inflicting either death or permanent disability. Indeed, surviving them can confer beneficial protection against future infections. Quarantine, a formal system of avoidance, may be an option in settled communities (e.g. the medieval lazar-houses for sufferers of leprosy), but is problematic in hunter-gatherer societies—being left alone during illness could be fatal.[16] Groups whose members were able to overcome their revulsion and care for their sick kin were therefore likely to gain three advantages in the longer term: not needlessly losing useful members, boosting group immunity, and increasing ingroup cohesion thanks to the strengthened bonds between survivors and their carers.[17] Liking, and especially love, can trump disgust.

Liking for others may arise from sexual attraction, from relationships of dependence, or from similarity.[18] Common ground may be recognized through conscious reflection, on the basis of conversations in which each party explores the other's view of important beliefs. Similarity may also be perceived much more swiftly and automatically through empathy: the remarkable human capacity to share experience.

Empathy comes in several flavours.[19] One, mediated by 'mirror neurons' (brain cells which are activated, for instance, not only when you move your hand but when you watch someone else move theirs), is motor empathy. Research suggests that our brains seem to use many of the same resources—similar areas of parietal and prefrontal cortex, for instance—when we make a movement as when we just think about it, so that our experience of watching someone else move has much in common with feeling ourselves make that movement.[20] This phenomenon can go all the way to the muscles. Highly empathetic individuals, the ones who twitch annoyingly during movies and come away with muscle pains and tension headaches, activate more of the neural patterns involved in the actual movements, making their imagined movements particularly detailed and realistic. But motor empathy can also occur at the much more abstract level of actions: that is, even for highly dissimilar *movements*, as long as the *action* is the same. When a cartoon boy, dog, ghost, or even fish—or a real animal—cringes in terror, so may we, although the cringing bodies move in quite different ways.

Motor empathy is intertwined, as the previous example suggests, with cognitive empathy, or theory of mind, and with emotional empathy. Theory of mind allows us to infer other people's beliefs, goals, and intentions from their behaviour. Empathy for emotions and the bodily sensations which ground them allows us to feel at least something of what others are feeling. Pain empathy, for example, seems to activate the same brain areas which are particularly involved in experiencing a person's own pain and negative emotions (such as the insula).[21] All three forms of empathy rely on the fundamental statistical facts underlying brain function: the correlations which ensure that similar events, on the whole, produce similar neural patterns. By imitating another person's 'events' (e.g. their gestures and facial expressions), you make your own patterns more similar to theirs. By using your own memories of similar situations to build expectations of what you would do in the situation you now see them in, you evoke the neural patterns which ground the feelings you had on those occasions. You may also evoke memories of their past behaviour to make your 'model' of their emotional reaction more realistic.

All three kinds of empathy are important for social interactions, especially when it comes to recognizing 'stop' signals. Cognitive and motor empathy allow such signals from victims to be detected by perpetrators of aggressive behaviour and compared against learned social norms ('this is how people signal submission') to deduce the victim's intentions.

Emotional empathy provides the motivation, in the form of unpleasant feelings triggered by the victim's pain and terror, to inhibit the aggression, as one's own pain and terror would.[22]

Empathy is a victim's last line of defence against cruelty. When social and legal conventions have failed, perpetrators can still be brought up short by the sudden recognition of shared humanity.[23] Even veterans can be affected. A notorious example from the Second World War occurred in the Ukrainian village of Byelaya Tserkov (now Bialacerkiew) in 1941, when the German army learned that about ninety Jewish children whose parents had been executed were being kept under guard in a house, without food or water, by Ukrainian militia-men. As the army's Catholic chaplain, who inspected the house, reported:

> The children lay or sat on the floor which was covered in their faeces. There were flies on the legs and abdomens of most of the children, some of whom were only half dressed. Some of the bigger children (two, three, four years old) were scratching the mortar from the wall and eating it. Two men, who looked like Jews, were trying to clean the rooms. The stench was terrible. The small children, especially those that were only a few months old, were crying and whimpering continuously. The visiting soldiers were shaken, as we were, by these unbelievable conditions and expressed their outrage over them.[24]

This heartbreaking story has no happy ending. Two days later the children were shot. SS-Obersturmführer August Häfner, who witnessed the executions, said later: 'The wailing was indescribable. I shall never forget the scene throughout my life. I find it very hard to bear. I particularly remember a small fair-haired girl who took me by the hand. She too was shot later...Many children were hit four or five times before they died.'[25]

People often succumb to the temptation to think of sadistic cruelty as morally worse (more evil) than callousness, because sadists fully intend to cause suffering (this is why Hannah Arendt's well-known remark about 'the banality of evil' caused such controversy).[26] In terms of what it does to victims, however, even callousness mitigated by concern can still be abominably cruel. In Byelaya Tserkov the Nazis' overarching goal was to eliminate Jews. Tormenting children, even Jewish children, was not a priority. Rather the reverse: as the chaplain noted in his report, soldiers in quarters near the house 'expressed extreme indignation over the conditions in which the children were being kept; in addition, one of them said that he himself had children at home'. At least some of the troops (empathy varies from person to person) were finding it hard not to see the children *as human children*, rather than, in the words of a less empathetic officer, as a Jewish 'brood' which 'had to be stamped out'. One lieutenant-colonel even complained, in a report to Field Marshal Walther von Reichenau, that 'measures against women and children were undertaken which in no way differ from atrocities carried out by the enemy', and that troop morale was suffering as a result. (Needless to say, this did not go down well with the field marshal.) Despite complaints that chaplains should 'limit themselves to the spiritual welfare of the soldiers', the Protestant and Catholic chaplains supplied bread and water to the children.[27]

This is not sadism but the callous pursuit of an organization's goal, defined loosely enough, in practice if not in principle, to leave room for minor acts of kindness. At least some soldiers felt empathy and concern for their victims, not delight in their misery. The German officers sought to downplay the impact on their personnel of the children's suffering (for example, by keeping troops away from the house and insisting that Ukrainians, rather than German soldiers, carry out the children's executions). They also tried to bolster their ideological justifications (for instance, by reiterating the urgency of the primary goal of Jewish extermination), and even to ease their victims' suffering to some extent. The lieutenant-colonel's report to Reichenau states bluntly that: 'Both infants and children should have been eliminated immediately in order to have avoided this inhuman agony,' thus reinterpreting callous murders as mercy killings.[28] Yet empathy and emotional discomfort failed to stop the massacre. Callous cruelty may seem less abhorrent than sadism, but the children of Byelaya Tserkov still suffered and died.

Why and when does being cruel encourage further cruelty?
One implication of the way brains work, noted in Chapter 4, is that activating a neural pathway makes it easier, next time round, to activate

related (overlapping) pathways. Talking about acting cruelly makes it easier to be cruel—unless one's talk incurs swift punishment. Acting out the otherizing ideas, especially in a group whose members compete for status and egg each other on, can push people into extreme otherization with remarkable speed.

In addition, brains habituate: a frequently repeated stimulus, behaviour, or emotion evokes less intense neuronal activity than its predecessors. More time spent in the earlier stages of otherization, especially when that involves consciously discussing and imagining more extreme stages, facilitates the progression to violent cruelty and makes it more palatable. Perpetrators who have not done much 'preparatory work' are liable, when confronted with actual atrocities, to experience strong, even disabling, emotions. An example is the German truck-driver confronted with the scene at Babi Yar, where over 33,000 Jews were killed in two days. 'I was so shocked by the terrible sight that I could not bear to look for long...I was so astonished and dazed by the sight of the twitching blood-smeared bodies that I could not properly register the details.'[29]

Veterans have had more practice in adjusting their thoughts and actions to reduce the emotional effects of their behaviour. Length of training matters, as military commanders are well aware. Some veterans may even be surprised by their own lack of feeling, like SS-Hauptscharführer Felix Landau, who after a morning's work—shooting Jews—muses in his diary: 'Strange, *I am completely unmoved. No pity, nothing.* That's the way it is and then it's all over.' This man was not incapable of emotion (he worried about his girlfriend), and he recognized that what he was doing conflicted with his own ideals. ('Isn't it strange, you love battle and then have to shoot defenceless people.') He even applied theory of mind to his victims—'What on earth is running through their minds during those moments? I think that each of them harbours a small hope that somehow he won't be shot'—and recalled his own feelings on facing death, years earlier: ' "So young and now it's all over." Those were my thoughts, then I pushed these feelings aside and in their place came a sense of defiance and the realization that my death would not have been in vain. And here I am today, a survivor standing in front of others in order to shoot them.'[30]

Perpetrators of callous cruelty need not always deny its impact on victims. Jews, seen through this SS officer's eyes, have thoughts and feelings; he predicts their behaviour as he would anyone else's (and is puzzled when they don't react as he expects). Although he recognizes their emotions, however, it does not seem to occur to him to share them. Jews are not suitable targets for emotional empathy; that would require acknowledging too much similarity. He has accepted Nazi justifications as legitimate, describing a Wehrmacht officer who suggests that Jews are under Wehrmacht protection as 'the worst kind of state enemy'. This is basic, ethnocentric morality at its most stark, because where his own group is concerned, his moral values are intact. He writes, apparently without irony: 'Beforehand we paid our respects to the murdered German airmen and Ukrainians. Eight hundred people were murdered here in Lemberg. The scum did not even draw the line at children.'[31] That scum, the Jewish-Bolshevik menace, is his otherized foe, deserving only retribution for its atrocities. Not all perpetrators internalize their governing ideologies to this extent. Those who do accept the beliefs as 'self' are more likely, later, to insist that their behaviour was justified.[32]

Why do people act cruelly when they know cruelty is morally wrong?

A central theme of this book, as of much recent work on perpetrator psychology, is that even the worst atrocities tend not to be committed by satanically wicked individuals dedicated to the glorification of evil. Far from it. The same individual may cuddle his own children in the intervals between killing other people's; may understand and try to live by the moral codes he grew up with; may, if he is religious, continue to consider himself an adherent; and may even make the same moral judgements when confronted with cruelty. The problem for his victims is that his morality seems not to apply to them.

What causes this tendency of moral prohibitions to 'lose their grip' on a person's behaviour? Imagine a soldier who has been ordered to kill a prisoner and who obeys despite knowing that the order is illegal. In neuronal terms the problem is one of insufficient activation; the patterns underlying his moral awareness do not have enough votes on the neuronal committees which take the decision to act. This may occur for two reasons. First, the patterns may be activated, but not strongly enough to inhibit the active, well-established neural circuitry laid down by military training. Moral patterns may be weakened by otherization, or may never have been very strong (for example, because the person never learned to accept these moral rules). They may become active too slowly to prevent the action from taking place, especially if it is easy and well-practised (pulling a trigger, for instance, as opposed to stabbing or beating the prisoner to death). Highly stressful situations, like being presented with an illegal order, can force people to rely on long-established patterns without allowing them the leisure to consider moral factors—unless those factors' underlying patterns are already strongly activated, which otherization will usually have ensured is not the case.

If otherization has been effective, however, then the soldier will have accepted that in this situation the normal moral rules do not apply. The prisoner is seen as not human enough (that is, insufficiently similar to the soldier) to deserve the social insurance provided by moral codes. In this second case the patterns underlying moral awareness may simply never be activated. They are still available, and at other times the soldier may demonstrate high standards of moral behaviour; but in this situation they are deemed irrelevant.

How does otherization achieve this remarkable narrowing of the moral horizons, suppressing not only empathy but sometimes people's extensive personal knowledge of their victims? Atrocities targeting strangers certainly happen (My Lai and Hiroshima are examples), but so do atrocities targeting the girl, boy, and family next door (as happened in Rwanda and Jedwabne, among many others).[33] The terrible irony is that otherization paints victims themselves as evil perpetrators: inhumanly depraved monsters who do not understand 'normal' morality and cannot be influenced by reason. From the Jewish blood-libel to a Serbian farmer's claim of having been sodomized by Kosovan Albanians using a bottle, no victims of atrocities fail to attract atrocity stories before they are attacked.[34]

Ludicrous as atrocity myths can be, they are widely believed at the time—and sometimes for generations, long after their adherents should

FIGURE 10. A notorious 1934 title page from the German newspaper *Der Stürmer*, a publication controlled by the Nazi and anti-Semite Julius Streicher, which describes allegations of ritual murder (the 'blood-libel') by Jews. Its message is clear: *Die Juden sind Unser Unglück!*—'The Jews are our misfortune!' The talk of a *Jüdischer Mordplan*—a Jewish plan to murder non-Jews—is designed to instil fear, hatred and anxiety. The otherizing language is reinforced by images which evoke disgust: ugly caricatures of Jews. The image of blood dripping into a vessel brings to mind the repellent idea of drinking blood. It also invokes Christian imagery; this propaganda wages its emotional war on several fronts.

have known better. There are still people out there who believe the *Protocols of the Elders of Zion* and the Holocaust deniers. One might call them gullible idiots, but this is not just stupidity (though that may be involved in some cases). It is world-shaping: the deliberate defence of strong beliefs against unwanted reality. We may laugh at their delusions (and it might not be a bad idea if we did), but that should not blind us to how dangerous world-shaping can be. As David Frankfurter comments in his book *Evil Incarnate*, 'in every one of the historical cases I address, it was the myth of evil conspiracy that mobilized people in large numbers to astounding acts of brutality against accused conspirators. That is, the real atrocities of history seem to take place *not* in the perverse ceremonies of some evil cult but rather in the course of *purging* such cults from the world.'[35]

How do ideologies and moral codes affect the prevalence of cruelty?

> *Noi vogliamo glorificare la guerra, sola igiene del mondo, il militarismo, il patriottismo, il gesto distruttore dei libertari, le belle idee per cui si muore.* [We want to glorify war, sole hygiene of the world, militarism, patriotism, the destructive action of anarchists, the beautiful ideas for which one dies.]
>
> (Filippo Marinetti, *Manifesto of Futurism*)

Cruelty involves *unjustified* harm-doing. Perpetrators, however, typically come well provided with justifications: otherizing stereotypes, false beliefs about the victim's power and hostility, and strong emotions to motivate aggressive 'self-defence'. Whether those emotions are stirred up by demagogues intent on grabbing power or are responses to uncontrollable stresses—like economic hardship, political impotence, loss, or harm by others—they aggravate internal conflict, inflaming the need for control and leaving the person looking for ways to ease that unignorable itch. Personality—that is to say, the accumulated interaction of past experience with genetic propensity—will shape both the search for options and the choice between them; but a person looking for answers, particularly a highly stressed person, will favour simple, easy (low-cost), and available solutions. Leaders, the 'political entrepreneurs' who stand to gain most from otherization, have huge responsibility here, because solutions from a trusted, powerful source often have their validity taken for granted (the assumption being that people who get things wrong tend not to be good at getting and keeping power).[36] Unfortunately, valid and feasible solutions to real-world problems are often anything but simple and low-cost.

Otherizing beliefs can fill the void between the supply of truth and the demand for security.[37] By blaming a specific human target, they present an uncontrollable problem as controllable, redirecting attention from difficult abstractions like 'the economy' to the personal sphere, the one humans feel they know best. They are familiar, fitting easily into pre-existing cultural patterns of belief. Better still, for those who seek to whip up intergroup hatred, they tap into ancient threat responses and suppress the empathy and moral awareness which might prevent hostility becoming cruelty. The otherizing beliefs which push people apart can stimulate the strong emotions associated with the 'threatening' outgroup (fear, anger, disgust, etc.) and the socially rewarding ingroup (love, pride, happiness, etc.). They can also channel strong feelings into the actions suggested by advocates of otherization. Those actions provide 'solutions' to the 'problem' presented by the target group. In a notorious example from Rwanda, the Hutu leader Leon Mugesera said of Tutsis: 'They belong in Ethiopia and we are going to find them a shortcut to get there by throwing them into the Nyabarongo River. I must insist on this point. We have to act. Wipe them all out!'[38] During the genocide, thousands of bodies were carried down Rwanda's rivers into Lake Victoria.

Empathy can be suppressed by activating incompatible emotions, or by presenting beliefs which challenge the link between empathy and the victim. Such beliefs ('Compassion is weakness!'; 'We must harden our hearts!') may explicitly contradict established mores which extol kindly behaviour. They may also focus on the victim's untrustworthy nature and past wickedness (real or invented), implying that any evidence of distress on the victim's part should be interpreted as faked, deserved, or both. Associating victims with negative emotions (especially disgust) and with qualities and activities despised in the perpetrators' culture (such as 'decadent' or 'effete' behaviour, and, of course, cruelty) makes them seem less similar, reducing empathy. By ignoring or downplaying victims' attempts to communicate with perpetrators (for example, by socially excluding them), and by severely punishing ingroup members who break ranks, otherization reduces the discomfort from signs of suffering humanity. That also makes victims seem less similar to perpetrators. By making victims seem more alike, and therefore likely to act cohesively, it also increases their apparent power, lending weight to the claims that they are to be feared.

Moral codes and ideologies can make cruelty more common within a society by making it acceptable behaviour. If leaders legitimize persecution by example and idolize the so-called 'warrior virtues' of ruthlessness and brutality (qualities many warriors do not find particularly admirable) while punishing expressions of empathy and kindness, they will shape a hyper-masculine 'honour' society whose codes of behaviour are rigidly enforced. Such societies typically feel threatened by powerful outside forces which may be perceived as seeking to deny them their rights (a status threat, provoking primarily anger), to change their moral essence (an identity threat, provoking disgust), and/or to destroy them altogether (an existential threat, provoking fear). To defend against these threats they prioritize group unity; hence the severe punishments for deviation.

As part of the process of distancing, a group may downplay beliefs it shares with its enemies. It may also emphasize claims which contradict the opposition's key ideals. These may or may not have been central to the group's belief system previously, but they can be made sacred if the group decides to adopt the additional dictum that challenges to them are not only wrong but morally unacceptable.[39] (Scouring the words of group leaders, past and present, for evidence to support this new interpretation is a marker of such transitions from belief to unquestioned dogma.) This sacralizing of beliefs heightens mutual otherization, reassuring group members that the Other is indeed alien. Examples of this wilful difference-building include the modern construction of 'Islamism', with its highly selective interpretations of Islamic religious texts, and Christian evangelicals' obsession with homosexuality.[40] Evangelicals do not obsessively emphasize 'traditional' families and gender roles because the New Testament is full of dire alarms about divorce, homosexuals, and career women; it isn't. Christianity's founder appears to have been more concerned about human cruelty: this, after all, is the man who told his followers to love one another, look after society's outcasts, and 'do unto others as you would have them do unto you'. Whatever your opinion of Christianity, these are morally admirable injunctions, still held up as ideals in the secular West.

And that is the point: these dicta do not face the challenges presented to less central evangelical beliefs (about the evils of divorce, sodomy, and uppity females) in a world where some ex-marrieds, gay people, and women proclaim their right to live as they choose while calling themselves Christians. The term 'right' is key. Gay activists and feminists who reject the traditional 'heteropatriarchy' do not see their choices as morally wrong. They say, in effect: 'My belief is a valued part of my identity, you cannot change it without making me not-me.' In doing so, they make their principles personal, putting names and faces to the symbolic threat and raising the stakes for those particular beliefs. The result? Onlookers see evangelicals, through the media's unsympathetic lens, as clinging to outdated, otherizing notions instead of promoting their faith's much more appealing moral core.

What motives drive people to be cruel?

Human behaviour has its wellsprings in the gap between how the world is and how it should be. World-shaping involves deliberate attempts to reduce that discrepancy by shifting the real world closer to the ideal. The results can be gorgeous or hideous, cathedrals or genocides; but successful world-shaping always involves a power relationship. Instead of patterns of neural activity changing to accommodate the aspect of reality which gave rise to them, reality itself becomes subordinate, adjusted to better match the patterns.

Exerting power is inherently rewarding, satisfying the needs for control and self-preservation—the evolutionary reasons, if you like, for acting. How those deep needs are interpreted at the psychological level will depend on the person's beliefs and circumstances. One person's fear can be another's exhilaration.[41] A man who feels threatened and lashes out in a frenzy may well be too busy defending himself, as he sees it, to have time to clarify the accompanying emotions. Later, if interpretation is needed, he may revisit his memories and label what he felt as fear, fury, horror, or whatever it may be. Which name he selects will depend on the physiological features of his experience—whether he felt sick, or his hands shook, or his heart was racing. But his choice will also depend on the circumstances and the reasons why he is seeking clarification. He knows from experience that fear is an acceptable justification in some situations, for instance, and fury in others. Indeed, he may consciously decide to call an emotion 'excitement' or 'terror' for the benefit of different audiences.

CONVENIENCE

> Our self-respect as a virile people obliges us to put down as soon as possible, by reason or by force, this handful of savages who destroy our wealth and prevent us from definitively occupying, in the name of law, progress and our own security, the richest and most fertile lands of the Republic
>
> (General Julio Roca (1843–1914), Argentinian minister
> for war, referring to the native Indians)

In a recent book on the psychology of mass killing, Daniel Chirot and Clark McCauley propose that perpetrators of large-scale atrocities have four main motives: convenience, revenge, simple fear, and fear of pollution, all of which may be present to varying extents in any given crime.[42] Convenience involves the pragmatic calculations of callous cruelty, which assess the costs and benefits of action for the perpetrator. The effects on victims are either ignored, downplayed, or justified, for example by using atrocity stories; if victims' suffering is recognized, it is not allowed to influence perpetrators' behaviour. Convenience is characterized by a focus on technical details and problem-solving (i.e. using pragmatic reasoning), backed up by selective use of ideological justifications, where required.

FEAR

> We have become like orphans in a banquet for the villains.
>
> (Ayman al-Zawahiri, *Knights under the Prophet's Banner*)

Costs and benefits, however, may incorporate emotional factors as well as considerations of effort, risk, and reward. The hunger for financial gain, the need to look good to one's peers, or the simple desire to bring the situation to a close may all feed into the decision to act. Sometimes, however, other factors may be drowned out by the strength of the emotions involved, especially when those emotions include the basic and intense reactions of fear.

When a target group is portrayed as highly dangerous, a realistic threat to self-preservation, fear can trigger actions unimaginable in calmer times.[43] In Rwanda many Hutus and Tutsis were terrified by the thought of the other faction being in power long before the genocide erupted in 1994, having lived through decades of ethnically driven bloodshed. Fear may motivate aggressive self-defence in response to existential threats from powerful opponents. In such extreme circumstances, most people feel that defensive violence is morally justifiable; aggressive, maybe, but not cruel. (Where the justification fails in the Rwandan case is in the assumption by militant Hutus that *every single one* of the estimated 800,000 Tutsis or moderate Hutu 'collaborators' who died was a lethal threat.)

Fear, however, evolved as a short-term defence against obvious, immediate threats. It is exhausting and, if sustained, extremely wearing, as its high demands for vigilance and readiness for action strain the heart and muscles and require intensive brain activity. Threats must be clear, realistic, and ongoing if the fear they evoke is not to seep away; and a constantly terrorized population risks collapsing into defeatism. For longer-lasting otherization such as that found in genocides, other motives—along with considerable organization—are also required.

ANGER

Wrath is cruel

(Proverbs 27: 4)

Revenge—what Romeo Dallaire calls 'the toxic pull of retribution'—falls into the category of anger responses.[44] In the perpetrator's eyes, the justification for what others see as cruelty is prior misbehaviour by the victim, deserving punishment. When otherization is in play, 'victim' can mean any member of the targeted group, alive or dead. Indeed, it is only within the context of otherization that statements such as 'They spied for our enemies during the war' or 'We'll never forget your crimes, colonialist oppressors' make any sense. The spies and oppressors in question are long gone, time-travel wasn't an option the last time I looked, and individuals born long after the events cannot sensibly be held responsible for them. Otherization, however, disregards individual agency, squeezing group members past and present into one homogenous mass, a crowd-entity in which agency and responsibility become assigned to the group, not the person. Any human constituent is interchangeable with any other in the group, and can thus be held responsible for any crimes committed by group members. And group membership can be made, by appropriate mythologizing, to stretch far back into history, strengthening the conception of an ancient menace and thus associating the target group—which itself may be of relatively recent origin—with older, often religious conceptions of evil.[45]

Essentialized as a single malevolent, powerful agent, the targeted group is seen as a threat to social order. Its behaviour challenges the power of the perpetrator, triggering the need for control and provoking anger, hatred, humiliation, and the urge for revenge. The perpetrator's reaction is seen as just that, a reaction, justified by the target group's power and hostility. In chimpanzee groups a rival who challenges the alpha male is put in his place with a snarl, slap, or fight if need be, until he produces the compensatory signals of appeasement which show that he accepts his attacker's superiority. Likewise for the dominant parties in human politics, where threatening groups must be suppressed until they stop threatening and start being properly subservient. If a group is not in fact very dangerous, then conspiracies must be uncovered linking it to a powerful external enemy. In the First World War the Turks claimed that the Armenians were working with the Russians; in the Second, the Nazis dreamed up an international Jewish-Bolshevik movement to bulk out their foe. If no obvious external threat presents itself, an inventive leader can usually exhume some past offences which, he can argue, did not receive appropriate compensation at the time. Slobodan Milosevic, lashing his followers to frenzy (in a speech in 1987) by ranting about the Battle of Kosovo (in 1389), is an example.[46]

DISGUST

> The life of this world is such that checking one group of people by another is the law of God, so that the earth may be cleansed of corruption.

> (Sayed Qutb, *Jihad in the Cause of God*)

Chirot and McCauley's fourth motive for mass killings is fear of pollution. Here the threat is neither to existence (as in fear), nor to social power, status, and convention (as in anger), nor to the possibility of achieving one's goals (as in convenience). Pollution threats affect identity. They may be lethal, but they do not so much *kill* the person as *change* them into something else. Fear and dread are certainly relevant to pollution, but the dominant emotion is surely disgust, for these are disgust-threats: subtle, not instantly lethal, hard or impossible to perceive directly, and preventable, with the right precautions.

Different kinds of pollution threats affect different aspects of identity. Physical integrity may be compromised, for example, by infection, with effects which can be truly horrifying. Gangrene, leprosy, or necrotizing fasciitis can turn living flesh into the stuff of nightmares. The Ebola virus can rupture internal organs and drown victims in their own blood.[47] Even the milder organisms dismissed by overworked doctors as 'just a virus' can transform us into dripping, sneezing, retching social pariahs. Dirt or colouration, masks and disguises can also be seen as changing identity by altering physical appearance, but here the identity moves from physical to symbolic, from varying appearance to veiling the self.

The fear of pollution referred to by Chirot and McCauley is largely concerned with symbolic threats to identity. When 'a particular group is so polluting that its very presence creates a mortal danger', the disgust response is an attempt at cleansing the self, whether by avoidance, expulsion, or elimination of the pollutant's ideas, customs, or carriers (target-group members).[48] Fear- and anger-threats are typically obvious and immediate, but disgust is different, making this kind of threat extremely relevant to our study. Cruelty, after all, is about *unjustified* harm: damaging people, often far less powerful than yourself, who do not deserve what you do to them and who pose no obvious threat.

Symbolic threats are often not obvious to third parties whose identity is not at risk from them. And the sources of such threats need have no actual power whatsoever, other than the power to express ideas.

Why do perpetrators so often seem disgusted by their victims?

> And they shall come thither, and they shall take away all the detestable things thereof and all the abominations thereof from thence.

(Ezekiel 11: 18)

For a political leader intent on stirring up hatred against a minority group, the target's low social status can pose a problem. How are one's followers to be convinced that a tiny and apparently harmless minority is in fact dangerously powerful and hell-bent on their destruction (especially if the minority has lived peaceably among them for years and has a generally positive reputation)? Any demagogue or tyrant deliberately setting out to promote otherization, whether because of genuine beliefs or political opportunism, will probably begin by selecting a group which already has a negative public image. Even so, the entrepreneur bent on violence has work to do, because violence is risky, unpleasant, and effortful.

In the Britain of the 1930s, thanks to centuries of Christian propaganda, Jews were one such unloved minority, subject to hostility from media such as the then pro-fascist *Daily Mail*, and physically attacked by rioters, most notoriously (but not exclusively) in London.[49] Today such overt anti-Semitism risks rapid condemnation, although subtler forms of anti-Jewish feeling, like many forms of racism, can still be found.[50] More blatant in today's anxious West are the otherizing simplifications currently reserved for those 'radical Islamist' Muslims who lurk undetected in our cities, plotting murder in the cause of making Britain a sharia state. Instead of world Jewry we have the jihad-ridden ummah, while the Jewish-Bolshevik conspiracy has given way to al-Qa'eda and its Middle Eastern backers.

The parallels are notable. Islamist extremists are presented as the violent tip of a monstrous iceberg of anti-Western feeling, driven by a primitive, barbaric ideology whose secretive cells manipulate young men into killing themselves and innocent bystanders—often betraying the country to which they should be grateful. This stereotype of the enemy as manipulative, treacherous, secretive, malevolent, powerful, and potentially lethal is remarkably similar to that traditionally pinned upon the Jews.

I am not saying, of course, that no Jew or Muslim has ever committed any atrocity. Which nation, race, or religion could make that claim about its members? There have been Jewish criminals, traitors and terrorists, just as there are indubitably people willing to murder 'in the name of Islam'. Stereotypes, however, draw their strength from their claim to be generic, to represent the group as a whole—and this is the problem. Most Muslim people, like most Jewish people, do not commit murder, are as trustworthy as members of other faiths, prefer their social order stable and peaceful, and pose no threat.

If someone's otherizing fantasies have a presentation problem, disgust is the ideal emotion to evoke. Why? Because disgust has three properties which make it particularly suitable to otherization, that denigration and distancing of other people which excludes them from our moral universe and makes atrocities seem both legitimate and essential.[51] First, the threat it reacts to is not immediate. You won't die from seeing a decomposing corpse, nor even necessarily from touching it; but chewing a piece of it could in time make you seriously ill. The absence of an immediate and obvious threat is therefore little hindrance for those wishing to stir up disgust among their followers, whereas scaring them rigid with empty terrors risks exposure, ridicule, and distrust. If someone is afraid of, say, Catholics, you can reassure them with stories of the kind and gentle Catholics you have known. If the person finds Catholics disgusting, on the other hand, your stories may be taken as evidence that you associate with Catholics, spreading the contagion of disgust to you.

(Did you feel revulsion at the three corpse-scenarios set out in the previous paragraph? If so, my guess is you found the third one the most repulsive. Disgust can be thought of as an alarm signal whose function is to stop you getting closer to the disgusting thing. Like those detectors in cars which bleep when you start reversing towards a wall, the signal gets stronger the closer you get to danger. Arousing disgust is thus an excellent way of maintaining distances—including distances between groups of people. This is why referring to someone as a 'cockroach' or a 'rat', calling them 'dirty', or even just suggesting they may be infectious can be such an effective method of otherization.)

The second property of disgust which makes it suitable as a social weapon is that disgust-threats themselves are invisible. Because human beings cannot detect pathogenic micro organisms like bacteria directly, they must rely on proxies such as the presence of dirt and decay. These are the stimuli which disgust us, because of their reliable past association with disease—but today that association is not inevitable. (That branch of modern art delighting in the excremental may look filthy, but most disgusting artworks are only symbolically, not actually, contaminating.[52]) Consequently, the link can be manipulated. Using disgusting metaphors (rats, cancers, plagues, insects, even toilets) to describe the target group can form symbolic links where none exist in reality.[53] As we saw in Chapter 6, disgust is particularly suitable for co-option by ideologues because ideas, like diseases, are intangible, easily spread, and sometimes lethal.

The third property of disgust which makes it suitable as a social weapon is that disgust-threats, unlike fear- and anger-threats, do not rely on the threat source's physical strength. Their power, the power of poison and infection, is sly and secret; potentially deadly in time, but helpless against us as long as we act to pre-empt them. Since using disgust-evoking metaphors tends to provoke less outrage among third parties than physical aggression, they can offer a low-risk method of promoting otherization. Its usefulness for targeting minority groups is also obvious. Disgust says: they may look harmless and trustworthy, but so can a meal which gives you violent food-poisoning. Lack of power, in other words, is no indication of inability to harm. Pre-emptive action is essential to prevent an infection which may prove incurable.

For minorities, or individuals, who live among the people learning to hate them, the problem is particularly acute. This is because a typical disgust response will involve withdrawal from the stimulus. One looks away, flinches, leaves the scene as fast as possible. When disgust is applied in the service of otherization, this avoidance often results in ghetto formation, as ostracized individuals join—or are forced—together. As social exclusion expands to include access to trade, work, and health-care, ghetto dwellers can become particularly vulnerable to poverty, poor health, and infectious disease, outbreaks of which are then used by their oppressors to vindicate and further inflame their otherizing language.

As otherization becomes more extreme, the 'problem' posed by the objects of disgust—and created by their persecutors—becomes more difficult to ignore. When the otherization is large-scale, as when a state is working up to genocide, the mass media provide an excellent means of spreading the necessary hyperbole. By making the targeted minority seem increasingly pestilent and in danger of escaping from its confines, they make the initial disgust response—avoidance—seem inadequate. The next, more effortful strategy is expulsion.

If, while walking down a street, you see the half-digested consequences of somebody's interaction with excess alcohol, you are likely to avert both your gaze and your involuntarily wrinkling nose, and hurry past. If someone throws up in your kitchen, however, avoidance by itself is not enough. Someone must clean up the mess, removing it to the location defined by custom as socially appropriate (e.g. the garbage can), and performing the appropriate cleansing rituals (e.g. spraying the area with disinfectant). Expulsion wraps the disgusting object in some sealed package, whose material acts as a safety barrier, before removing it; this prevents actual contact. These precautions, like the cleansing rituals, may or may not be practically effective at killing pathogens, but they symbolize the restoration of hygienic standards, of an accepted social order which does not include vomit-smeared kitchens.

In other words, disgust, as well as protecting against infection, prompts action to restore the social order.[54] If a minority is seen as violating that social order, and avoidance appears ineffective or impractical, the obvious next step is expulsion, often using specialized personnel (for example, soldiers or paramilitaries) to form the social equivalent of a safety barrier. The Nazi plans to send Jews first to the East and then to Madagascar were attempts to explore this second strategy.[55] Jews were separated from a population as a whole, isolated in ghettos, and deported from Germany in the custody of trained guards.

If expulsion fails, of course, one is left with a third strategy: elimination. Applied to either people or objects, this is a last-resort gambit: it takes more effort and involves the risk of longer contact. Destroying the object not only consumes energy but may risk leakages of dangerous fluids or gases as the object's structure breaks down. Nevertheless, it transforms disgusting material into safer—that is, less contaminating—stuff, like ashes. If the costs of avoidance and expulsion are much higher than the costs of elimination, therefore, the latter may be used.

Mass killings occur when avoidance is not a realistic option. German society in the 1920s and 1930s was never going to remove itself en masse from German territory to avoid the Jews located there. Disgust responses thus tend to begin with expulsion–segregation, ethnic cleansing, forced migration, and resettlement—proceeding to elimination once expulsion is perceived as having failed. As the pressures and privations of war began to bite, as the implausibility of the Madagascan alternative became obvious, and as German commanders in the East kept complaining about the logistical nightmares arising from the numbers of Jews they had rounded up, elimination became the approach of choice for the Nazi leadership.

Disgust provokes the pragmatic responses of callous cruelty. If those responses are not checked, a person otherized by disgust can come to be seen as contaminating waste matter, to be removed or destroyed. As the derogatory metaphors take hold of thought and action, they trigger thoughts of habitual responses: washing and disinfecting, burning, burial. In extreme otherization those thoughts can spill over into actions, with devastating results—like the bodies of Tutsi shoved into rivers in Rwanda, the Jewish corpses buried at Babi Yar, or the dead burned at Auschwitz.[56]

Summary and conclusions

Callousness to other people varies from person to person. Individual differences in cognitive, motor, and emotional empathy, stress levels, suggestibility to pressure from peers and leaders, the capacity—and time—to pull apart otherizing arguments and seek out alternative solutions to problems, and personal experience—these are just some of the factors combining to make some people more prone than others to act callously. Whatever the causes, callous cruelty always begins with otherization and with more or less flimsy justifications for damage done or planned.

Sometimes the victim's existence may barely be recognized—an example is the drug addict whose need for a fix drowns out countervailing factors. Sometimes the justifications are genuine: not all conflicts are self-inflicted, and not all victims are innocent of cruelty themselves. In between lies a horde of unflattering portraits of the human species: frightened, lazy, stupid, venal, deluded, vengeful, or just plain indifferent, looking out for themselves and their own, looking for someone to blame for their unfulfilled expectations, or just looking for some way to feel clean again, and safe, and suitably respected.

Most cruelty does not kill—which is just as well, considering how commonly we use it. Someone must be doing all the bullying we hear about, tormenting the thousands of animals rescued each year, abusing partners and children and elderly relatives, or even just screaming at babies or sneering at fat people or crossing the road to avoid a tramp (we call them 'homeless people' nowadays, but whatever the label they still feel the hurt and humiliation of ostracism). None of this behaviour is strictly necessary. It is the easiest way the perpetrators can think of to make themselves feel better, temporarily; but since when did human beings have an automatic right to feel better? We don't, of course, whatever the US Constitution may declare. But once we evolved ways of feeling good at all, we were stuck with the motives that drive us to try for improvement.

The extreme cruelty of atrocities takes minor cruelty as its raw material and uses otherization to amplify it almost beyond recognition. Less violent viciousness, such as verbal bullying, is deliberate in the sense that the behaviour is voluntary, however instinctive or ill-thought-out it may feel to the perpetrator. Extreme cruelty, such as the torture of individuals or the massacre of groups, is deliberate in the sense of having been planned, sometimes for years. It demands organization to equip and train the perpetrators, to manufacture or enhance the threats to identity and survival which will be used to justify the atrocity, and to spread the propaganda which will turn the victims into villains. Justifications must be provided, a sense of urgency whipped up, critics silenced, and a clear and simple vision provided for followers. The political entrepreneurs who lead their followers down the path of otherization always attempt to instil a sense of claustrophobic necessity: 'these actions are vital, we have no alternative.' If that is true, it is because they have succeeded in closing off all practical alternatives. The momentum has become unstoppable, and only third-party intervention can save the victims.

Outgroups become targets for otherization because they pose a genuine threat, or because they can usefully be made a proxy threat, a soluble problem replacing all the insoluble complexities about which no single human can do very much. If they, or their allies, are too powerful to attack, or if the society's checks and balances provide a less favourable climate for otherization, then the hostility may be restrained, as hostility to Jews is currently restrained in Britain. One way to do this is to have a thoroughly disunited society, with power distributed among many competing groups, but that can be unnervingly unstable, and those currently in power are unlikely to approve of such arrangements. This is why modern democracy, even in the West, is strictly limited.

When the powerful come to believe, for whatever reason, that the only solution to a problem is to remove a particular group of people, the logic of otherization becomes inexorable. The only way to be absolutely sure that an outgroup will not remain a threat is to eliminate it. Anger and fear will help to trigger reactive aggression, if the threat is credible (or can be made to seem so), but they risk being propitiated once the victims' suffering becomes apparent. Sustained and callous cruelty is required, not instinctive lashing-out, and for that disgust is needed. So the leaders will reach for their metaphors: cockroaches, pestilence, cancer, garbage, dirt, excreta, rottenness. At which point, unless someone intervenes, people become much more likely to die in large numbers.

In this chapter I have focused on the negative emotions which accompany cruelty. But cruelty also brings positive emotions. As well as the direct benefits of callously ignoring other people's feelings, cruelty to outsiders can reinforce the bonds between ingroup members. It can bring substantial rewards, which is why callousness is part of basic human morality, though constructed moral codes may frown upon it. If praised as courage or self-sacrifice, cruelty can also be a source of powerful social reward: affiliation, the feeling of belonging, which at its strongest becomes the overwhelming power of love. Love, like hate, is a potent force for evil. It can turn the urge for revenge into obsession, transform fear for oneself into fear for one's family, friends, or even one's country, and by making the ingroup more valuable, more sacred, sharpen the knife edge of revulsion which leads to atrocities.

Love, furthermore, is not the only positive emotion. Sometimes cruelty seems to offer enjoyable feelings which have little or nothing to do with praise for protecting ingroup members, and much more to do with the agony inflicted on the victim. This is sadism, and like callousness, our reactions to it can often be disturbingly ambiguous. Reviled throughout history, yet a staple of light entertainment then and now, sadistic cruelty is the topic of our next chapter.

Chapter 8

Why does sadism exist?

The more we saw people die, the less we thought about their lives, the less we talked about their deaths. And the more we get used to enjoying it.

(Fulgence Bunani, a perpetrator in the Rwandon genocide,

???)

M ost cruelty is about callousness. It may involve failing to realize, ignoring, or deliberately downplaying the harmful effects of cruel behaviour on victims, or it may involve the belief that the harm is justified: fair punishment rather than unfair vindictiveness. Perpetrators of callous cruelty may acknowledge that one of their goals is to inflict suffering, but they see that suffering as instrumental, a means to an end. Sometimes, however, the means becomes an end in itself. The victim's anguish becomes rewarding and desirable, and creating or prolonging suffering becomes the perpetrator's primary aim.

I should note that sadism as discussed here is not necessarily the same as the sadistic personality disorder or sexual sadism of psychiatric literature.[1] These rare, much-debated, and ill-defined clinical conditions are poorly understood, in part because of the paucity of research.[2] Clinical studies of sexual sadism are few, especially outside the psychoanalytic literature.[3] Sadism as it has come to the attention of psychiatry is overwhelmingly sexual, homicidal, or both; whereas sadism as defined in this book need involve neither lust nor murder, as long as the victim's suffering is enjoyed for its own sake rather than seen as a step towards some other goal.[4] Moreover, some sadism (in war, for instance, or in the psychology laboratory) appears to be perpetrated by people without obvious pathology.[5] For all these reasons, the focus of this chapter will be on sadism as a behaviour rather than as a psychiatric disorder.

Sadism may be arcane in the medical literature, yet it is central to the everyday concept of cruelty. The delight in hurting is what makes cruelty not only evil but the defining characteristic of human evil. In this chapter I will look at that association, asking why, if sadism is so detested, it permeates our entertainment and haunts our news media. What conditions give rise to sadism? What motives drive it? Given how much atrocious cruelty we see on our screens, is everyone a secret sadist? In a subject defined by the abnormal, let us begin by reaching for normality. How widespread is sadistic cruelty in everyday life?

Sadism is rarer than it seems

Sadism is much rarer than callousness. With respect to sexual sadism, estimates from the United States and Canada suggest that only a few per cent of homicides are sexually motivated; the prevalence of sadism as a motive is not known.[6] In the general population, a benchmark study of sexual behaviour found that 12 per cent of women and 22 per cent of men experienced some level of sexual arousal in response to sado-masochistic stories. This sounds dramatic, but the numbers should be treated with caution since they lump together sadism and masochism.[7] Moreover, while acts of BDSM (bondage, discipline, sadism, masochism) appear to be a common part of the human sexual repertoire, most are mild and inflict little or no pain. Exerting control and achieving sexual arousal, not causing suffering, seem to be priorities. Interactions among serious BDSM practitioners (a much smaller community) are typically highly regulated, with pre-agreed 'stop' signals, conventions as to which behaviours are acceptable, and so on. This may be a form of 'vicarious cruelty' (of which more below): acts of violence and compulsion which allow the dominant partner to imagine that he or she is being cruel. The use of a willing victim reduces risks and costs for both parties. BDSM is thus not true cruelty—except, of course, if the dominant partner is over-enthusiastic.

Cruelty does not require the rituals of BDSM. It may not involve abuses of the flesh at all. Physical and psychological abuse can devastate victims without approaching the gory extremes of torturers or serial killers, in workplaces, schools, care-homes, and especially families. Whether all this viciousness is genuinely sadistic, however, is debatable. As with extreme cruelty, less spectacular cases may be driven by motives like financial gain, revenge, and fear. World-shaping may also be involved, for example when lower-status employees, children, clients, or family members do not behave as the perpetrator expects them to behave. As with the case of the SS man and the baby described in Chapter 6, response and threat are closely interwoven. A carer who thinks her Alzheimer patients are disgusting may abuse them by leaving them uncleaned; one who thinks they are aggressive may torment or physically abuse them; and so on. All this is undoubtedly cruel, but unless the primary aim is delight in causing pain, it is not sadistic.

And yet the human imagination is steeped in sadism. Horrific assaults and ingeniously unpleasant murders are abundantly available in mainstream fiction, whether that fiction sits decorously on your bookshelf or pours from the nearest screen. Savage tortures are meted out to heroes and villains, women are raped and murdered, civilians ripped to pieces by aliens, CGI armies slaughtered and virtual cities razed to the ground. Far worse is available on the Internet, where you can, if you wish, download real live videos of real live innocents being bombed and beaten, children being tortured, hostages being brutalized and killed.[8]

Watching someone suffer can be painful. Watching someone weep, or bleed, can be disgusting. Watching someone die can be terrifying. Watching someone defecate is also deeply unpleasant for most people; yet we treat the latter very differently. Why is it unacceptable to film someone committing coprophagy, which is unusual but non-violent, and yet common to film a person being shot dead, eaten alive, or blown to bits? As we saw in Chapter 2, the intense pressures set up by Philip Zimbardo's realistic prison role-play, let alone the compulsions of war, may have the power to foster cruelty. Yet what is it—in the freedom and comfort of our imaginations—that makes us place such emphasis on sadism?

Sadism and evil

> Indeed, we ourselves very much crave a rhetoric of evil—
> and the certainty that follows the deployment of a word like
> 'evil,' signifying what is absolutely inhuman, beyond the pale
> of comprehensible behavior, and of a nature that transcends
> the individual atrocity.
>
> (David Frankfurter, *Evil Incarnate*)

Calling somebody 'sadistic' is not a value-neutral judgement in common parlance. It implies condemnation of the person's behaviour, setting it within the context of a moral code accepted, it is assumed, by both speaker and hearers. But which code? As we saw in Chapter 1, evolution has provided us with basic, group-centred morality and with the capacity to build more complex moral systems. In belief systems nurtured in arduous circumstances (like the fight for survival), basic and constructed morality may show considerable overlap, while societies less plagued by war and the struggle for resources can diversify moral codes far beyond their basic origins. One form of constructed morality found in Western societies today, for example, is the medical view of sadism as a pathology, a sickness of the brain. Sadists are ill and need treatment, not punishment.[9]

Basic morality, unlike its constructed forms, is clear and condemna-tory: sadism is the vilest of offences. It provokes revulsion, outrage, and punishment: the traditional disgust-threat responses of avoidance, expulsion, or destruction of the offender. Sadists, like those who commit the worst atrocities of callous cruelty, are frequently described as 'evil'. As we saw in the Introduction, this opens the way for vigorous collective action against anyone accused of sadistic cruelty.

Callousness, the pursuit of cruelty for other reasons, can be a sign of strength when, directed against outsiders, it brings home benefits. In the blind scales of natural selection, behaviour weighs more heavily if it helps to boost gene survival, not moral well-being. Basic morality considers it acceptable for men, the primary defenders in our ancestral groups, to boast of their callousness, compete to outdo each other in warlike behaviour, and show a keen interest in violence from a young age—as long as this aggression is socially controlled and directed against non-members only. Deviation from this 'honour code' is punished. Being too soft on the enemy, displaying cowardice, or showing weakness when faced with callous cruelty can incur vicious social penalties, but a young male also soon learns that unjustified violence against females, children, or the elderly is likewise unacceptable (although what a particular culture considers 'justified' varies widely).

The social status of the resulting warrior code varies across and within societies (it remains much stronger in the southern states of the United States than in the northern states, for instance).[10] Even when officially frowned upon, it remains an influential moral system: a standard framework of the movies and books and tabloids which show us the people we would like to be. If no virile hero ever meted out rough justice to his dastardly foes there would be no popular film industry worth mentioning. Sadism is different. Even when directed against contemptible villains it turns our stomach, not necessarily because of what is done (sadism need not involve blood and guts, or indeed any physical damage), but because we have evolved to sicken at the sight of such naked and gratuitous cruelty. Why? Being good Darwinians, we must presume that the reaction conferred some evolutionary benefit. If earlier humans who punished signs of sadism in others were more successful players of the gene game than their more lenient contemporaries, our visceral response to sadism, and the ease with which it grabs our attention, might both derive from that selection pressure.

We cannot return to those ancient days, so any speculation must remain just that; but one can see why sadism might come to be so vigorously suppressed. Even when its initial targets are outsiders, the pursuit of cruelty for cruelty's sake is a threat to ingroup members. What happens when no enemies are available? What is to prevent the sadist targeting a spouse, child, or neighbour instead? Normally, moral prohibitions, backed up by the threat of community sanctions, would suffice, but communities know that strong desires can override both moral rules and sensible self-interest. As the desire to hurt comes to dominate, other incentives lose their grip. The sadist, like the drug addict or the fanatic, is no longer interested in the greater good, or goodwill, of the group, in social benefits, or even in the urge to propagate their DNA. Their hunger requires satisfaction, whatever the cost in damaged friends and relatives. Since they cannot be trusted to spare even their closest kin, if no alternative victim is available, they must be subjected to the far harsher sanctions of physical control, and, in extreme cases, ostracism or expulsion from the group. And because of the human tendency for mild bad behaviour to lay the groundwork for worse, groups which punish signs of sadism early on have more chance of preventing it from becoming dangerous. We can begin to see how powerful moral judgements could evolve.

Sadism, like all cruelty, involves unequal power relationships (necessarily, since the perpetrator must be able to act against the victim), and in sadism the victim's powerlessness can be an enormous source of reward for the perpetrator. Groups which refused to tolerate sadistic behaviour thereby protected their most defenceless members, including the children who comprised their investment in the future—after all, young people exposed to cruelty may either die or grow up to be dysfunctional parents.[11] Punishing and abhorring sadism makes excellent evolutionary sense.

Why does sadism exist?

> Extermination is one thing, but there is no need to torture your victims beforehand.
>
> (Eugen Horak, interpreter at Auschwitz, cited in Richard Overy's
> *Interrogations*)

What this Darwinian just-so story has not yet explained, however, is how sadism could have originated and survived in such a hostile climate. Indeed, there is a wider question: *can* evolution provide an answer to the conundrum of sadism?[12] Callousness, like the moral judgement which punishes sadistic cruelty, is a natural consequence of evolution's gift: our basic morals. Yet inflicting suffering for suffering's sake seems as much a paradigmatic case of non-evolutionary behaviour as, say, contraception. Killing a stranger to prevent his genes competing with yours is one thing; but what is the point, in Darwinian terms, of torturing him first? Why not simply bash his head in and be done?[13]

THE COSTS OF CRUELTY

To human beings, used to having to pay for and manage resources, natural selection can seem highly inefficient. This is the system, after all, which came up with the dodo. It tends not to tidy up after its experiments, leaving useless and sometimes dangerous bits and pieces lying around (as anyone who has had appendicitis knows). It creates extravagantly bizarre creatures which promptly cease to exist at the drop of a habitat (or else cling on by a metaphorical whisker, thanks to the frantic efforts of conservationists). It sends trillions of seeds and sperms and tiny babies into a world in which only a few can survive. Why? Because in the past these were strategies that worked.

At the individual level, the very extravagance of natural selection makes energy conservation a priority for any organism. Resources are often scarce, and other hungry eaters all too plentiful. In hard times, prodigals will be less likely to make it through than tightfists. The fluid grace of a loping wolf, leaping fish, or soaring eagle, each of which wastes admirably little energy, reflects millennia of gradual optimization, as efficient ancestors generated more descendants than their less thrifty rivals. Even human beings, who gaze in awe and envy at the matchless ease with which animals fit their environments, are adept at cutting corners, making life less arduous, and seeking out the path of least resistance. Laziness may traditionally be a sin, but energy conservation is a deep-woven part of the human fibre (wasting other people's resources is a different matter). Isn't that why we use contraception—in order to take our pleasures without the inconvenience of being lumbered with expensive, time-consuming infants?

To be sadistic is to veer off the path of least resistance, wasting resources on an activity which brings no obvious benefit. Serial killers, those archetypal sadists, can spend inordinate amounts of time and energy planning their crimes, then catching, restraining, and carefully damaging the victim. Given that the outcome is at most equivalent to murder—one competitor out of the reproductive game—why not simply kill quickly and save the surplus energy for use elsewhere? The logic of evolutionary theory is that '[a]ny such pattern of conspicuous outlay demands an accounting'.[14] Not only is sadism wasteful, it can be dangerous, risking damage from the victim, reprisals from their allies, and even disgusted ostracism from one's own kin.

We often think of sadists as loners, predators bolstered by the techno-logical supremacy of wielding a weapon and targeting a helpless victim. Yet if sadism is evolutionarily ancient it must have been fostered in a world very different from ours. Our early ancestors are thought to have lived in small, close-kin groups, at risk from natural hazards, predators, poisons, diseases, and other groups of humans. Females and children, a precious resource, were unlikely to be left undefended. In such a world lone stalkers would have run a higher risk of death than males who stayed with the group, since the latter could share food and sentry duty, or call for help when facing powerful predators.

Social life clearly worked, from the genes' perspective, given its centrality in human existence today and humanity's success at controlling other species (insects and microorganisms aside). Yet sadism is about as antisocial as behaviour can be, and unlike other antisocial behaviour, such as cheating on partners or free-riding, its benefits for reproductive fitness are not immediately apparent. Of course, diagnosing cruelty requires access to motives—difficult from prehistoric data, although there is considerable evidence of intergroup violence—so we cannot be certain that sadism afflicted our early ancestors.[15] But we know that as far back as the classical world excessive cruelty was a cause for complaint, evoking horror like that we feel today.[16] Sadism is not simply an Enlightenment pathology invented by a French aristocrat with a dodgy sense of humour. It has been part of the human repertoire for long enough to have been subject to the forces of cultural, if not genetic, evolution. It lies at the heart of our concept of evil—hardly a marginal position in the human imagination—and it pervades our choice of entertainment. Why should imagined, vicarious cruelty be so much more popular than the real thing? Many of us have violent, even homicidal, thoughts, though few of us act on them.[17] Yet if sadism is so harmful and unpleasant, one might expect it to have been eliminated from the human repertoire long ago.

Why has this not happened? Is it because a taste for sadism is pro-mulgated by dominant cultural forces, such as religion? Are the great and the good really feeding us cruelty for the purposes of social control: offering the masses attention-grabbing outlets for emotions which might otherwise fuel revolt? Alternatively, is the market in sadistic dreams a result of some vile flaw in human nature, generally kept in check by cowardice, but revealing itself in fantasies and, occasionally, actions? If so, does that flaw affect all of us? Or, in an age when character traits are treated with pills, can we see sadism as just another pathology, afflicting only a few unfortunate souls?

SADISM AS A PATHOLOGY

Could sadism be pathological, the result of some internal malfunction, like appendicitis? Researchers have various proposals for what that malfunction might be, including structural problems, such as brain deficits ('his insula's too small, Mrs Smith, that's why he eviscerated your cat'), and functional problems, such as abnormalities in the way brain regions talk to each other. Structural problems, notably physical damage to the frontal lobes, have long been linked to antisocial consequences.[18] Functional problems can arise as a consequence of structural damage, as the brain adapts to work around the damaged areas. They may also reflect abnormal environments, as the developing brain is altered by the impact of, say, serious abuse in childhood. For instance, sexual abuse may cause the victim to associate physical pain with sexual arousal. Depending on the child and the circumstances of the abuse, the child's reaction may involve accepting the association, leading them to behave sadistically or masochistically in later life, or guiltily rejecting it, leading to denial, depression, and/or post-traumatic stress.[19]

Individuals, whether perpetrators or victims of sadism (or indeed both), are normally only assessed in a clinical setting. We thus have little information about how their brains functioned prior to the experience of cruelty, and little knowledge of the causal pathways (and obtaining such data is ethically problematic). As an example, suppose that sadistic child abusers who were themselves abused as children are shown to have abnormally functioning prefrontal and temporal areas of cortex. Even if the research is statistically powerful, reputable, and replicated, it tells us very little by itself. The observed brain dysfunction could in principle be due to genetic factors, to damage inflicted early in development (perhaps even in the womb), to the specific effects of the abuse, to some later occurrence, or even to some factor coincidental with the abuse (e.g. poor nutrition) which may or may not have been caused by the original abuser.[20] Until these problems are overcome by obtaining data from early in life and in large numbers of people (how long before regular brain scans are as much a part of general health-care as blood-pressure tests?) the rarity of sadistic behaviour and the difficulties involved in studying it will continue to present formidable problems for those researching pathological models of sadism.[21]

Some sadism is almost certainly due to serious brain damage or dysfunction. Yet many atrocious tortures are carried out by people in specific situations who may otherwise show no signs of cruelty. The brain may function abnormally during these situations, but if this is pathology it is a deficit that comes and goes in response to social cues, and one to which even healthy people are susceptible. Is every perpetrator of extreme cruelty in an unusual brain state at the time? Possibly; we don't know. Can we describe that state as abnormal? That is a matter of semantics; is hunger abnormal? Cruelty is something done by people with and without dysfunctional brains. In general, healthy people must be immersed in unhealthy situations, like war, before they will escalate everyday cruelty into its extreme and lethal variants. Less healthy people may express their cruelty more readily in normal life.

The capacity to be cruel is part of us. We are not justified in otherizing it, except in certain rare cases, as some kind of disease, but nor should we conclude that every human is innately cruel. For most of us, most of the time, sadism is rare and altruism hardly unknown. Tendencies to be cruel, at times, we undoubtedly possess, but tendencies are not the same as the instinctive relishing of other people's pain. Very few of us are cruel in the way that we are hungry. Yet just as lack of food can drive the starving to cannibalism, so conditions which encourage cruelty can turn good citizens into sadistic killers.

CRUELTY AND SOCIAL CONTROL

What of claims that sadism is promoted by culturally dominant powers? Much has been said about the importance of intellectual and political leadership and state connivance in the large-scale atrocities where sadism is apparent.[22] Another commonly fingered cultural culprit is religion. Christianity, with its incalculable influence on Western thought, is the obvious example, given the act of appalling cruelty set at the core of its belief system (a phenomenon found in many religions). Even apart from Jesus's crucifixion, scenarios of ghastly cruelty have been there for anyone who could read since the development of print brought religious texts to a mass market (in Foxe's *Book of Martyrs*, for instance).[23] Long before that, religious art made extreme cruelty visually familiar to people who might never have witnessed it directly.

Pace comments from the likes of Friedrich Nietzsche, however, this emphasis on cruelty cannot be taken as evidence of Christians' uniquely unhealthy obsession with pain and suffering.[24] Victimhood may have been democratized by the social pressures of secularization, leading to fewer musings on saints' tortures and the crucifixion and more on the sufferings of ordinary people caught up in terrible events. Yet this says more about shifting patterns of social control than it does about cruelty; the presumed decline of faith has not stopped complaints about how vicious modern Western entertainment, media, and culture are at times. Religion's power may ebb and flow, but cruelty was a feature of humanity's fantasies long before either Hollywood or Jesus. It just happens to be the case that for most of our history, much of the symbolic content of human imaginations was provided by religion, so vicarious cruelty mainly occurred in a religious context.

Religion inevitably addresses pain and suffering. That is in large part what it is for. Furthermore, imagined and actual cruelty are easily found in cultures which disavow religion (Cambodia under the Khmer Rouge is an extreme case). Cruelty features promiscuously in the fantasies of today's secular Europeans and their more devout American cousins. Its popularity may vary across groups, just as it varies from person to person, but it is not restricted by religion or culture, either to the Abrahamic faiths or to 'exotic' cultures like the Aztecs and the Mongols. The myths of ancient Greece—that culture traditionally praised as the fount of Western civilization, democracy, and reason—could be appallingly cruel: think of the legend of Prometheus, punished for helping humankind by being chained to a rock and having his liver torn out by a hungry vulture.[25] As executions go, this is cruel and unusual, inhuman and degrading, but the story's sadistic genius is that Prometheus is not executed. He survives, his liver regrows, and the bird returns...

Religion, like other belief systems, can channel and encourage cruelty, real and imagined (as well as providing the impetus to resist it in some cases). So can disputes over territory, leaving young people with not enough to do, and being a parent. Tempting though it might be to blame religion, eradicating religious belief, could that be done, would have little or no effect on human cruelty. There are too many other reasons to be cruel.

As with most nature–nurture discussions, the reality is that both come into play. Culture mediates the paths our imagination travels. Evolved human nature provides the deep reasons why we find such dreams enjoyable. Together they give us a third response to the problem of sadism: the uncomfortable yet obvious thought that people are cruel when they find cruelty rewarding. Our capacity to savour pleasures is so formidable that it can overcome even the horrors of committing atrocities. Sadists emerge because cruelty can be fun.

THE GLAMOUR OF SADISM

If cruelty can be enjoyable, our taste for vicarious cruelty makes sense: it offers us pleasures without their unpleasant consequences. In the real world victims are noisy and messy and have to be disposed of; they can be dangerous, and they carry the risk of social retaliation. In fantasy, they usefully evaporate as soon as the consumer loses interest. But why would we want to savour any kind of cruelty? If imaginary viciousness is rewarding, what are its rewards?

One reason for the glamour of sadism has to do with cruelty's emotive power. Stimuli which produce strong negative emotions are generally more salient (have more impact and attention-grabbing capacity) than those which produce more positive or less intense emotions. Cruelty is highly emotive, both because of the stimuli involved—blood, screams, and suchlike—and because we are highly attuned to signals of social power and status. People who break moral rules, in reality as well as in fiction, are perceived as powerful, especially if they are charismatic, likeable, and appear to get away with their transgressions. This makes them inherently interesting. We can identify with them, albeit partially and transiently, and thereby enjoy a vicarious sense of power, enhancing our sense of control without having to face the dreadful after-effects of real-life cruelty. (Technologies of war similarly place distance between the action and its consequences, which has helped to make war more enjoyable for those involved, particularly if they have moral qualms about the cause for which they are fighting.[26]) It is no coincidence that most movie violence minimizes the impact on the victim and focuses on the power of the perpetrator, thereby assisting our imaginations to do the same. Action movies like the *Indiana Jones* series, in which large numbers of one-dimensional villains and bit-part players get eradicated, are of this type. Moral inhibitions which would normally be active are themselves inhibited, since we know the violence is fictional. We like the perpetrator and we know that his behaviour is not our responsibility (since our identification with him is only partial).

Other perpetrators are portrayed as cartoonishly brutal sadists who get their well-deserved come-uppance. This allows us to sample sadism in a controlled, not too upsetting, and clearly moral fashion (in the sense of basic, us–them morality; human rights and turning the other cheek are irrelevant). By exaggerating the nastiness of the villain's behaviour, often to the extent of making him or her a caricature, the fiction reassures its audience that evil people are easy to spot, quite different from us, and inescapably headed for well-deserved punishment.

In the *James Bond* movies, for instance, the villain is typically marked as a sadistic bad guy early on by disposing of a good guy in an ingeniously gratuitous fashion: dropping his victim from an airship, or into a giant mincer, or some such. These revolting deaths are lapped up by the audience; yet, *pace* theories of the importance of imitation in human culture, viewers tend not to leave the movie intent on pushing the nuisance in their lives into an industrial shredder. People who do that, the film tells them, will get it done to them. Bond villains meet ends as sticky as or stickier than the ones they doled out, and usually gruesomely appropriate. He who drops people out of airships falls to his death; he who employed the mincer ends up minced.

One further aspect of fictional cruelty is worth mentioning. Portraying exceptional cruelty can sometimes involve such extreme otherization that the cruel individual is less a person than a form of evil incarnate: as inexplicable, powerful, and terrifying as a force of nature. The character of Hannibal Lecter, the psychopath with a taste for human flesh introduced to us by Thomas Harris and Hollywood, is an example.[27] In such cases, the viewer is encouraged *not* to identify with the villain. This is achieved in part by making the latter's motives opaque and his personality inscrutable, in part by the extremity of his behaviour. Not many people even fantasize about digging out mouthfuls of somebody's brain with a spoon, as Lecter does, let alone do it. Like an earthquake, he seems to strike at random (at least until we work out the plot), a grinning monster sadistically wreaking havoc.

But there is one person he never strikes down: the sympathetic heroine with whom the viewer is to identify. Much of the story is about her attempts to determine his motives for action: to impose a pattern, in other words, on his chaos, making him tamer and more predictable. That she is somehow special to him is obvious. Other people die, more or less revoltingly, at Lecter's hands; but not Clarice Starling; in the movie *Hannibal* he even goes so far as to save her life. The message, which is the message we most love to take from fiction, is precisely that sense of being special. Cruelty may be raging all around you, it whispers, but you are important. You matter—even to wicked people, tempestuous nature, or the supernatural force of evil itself. Others may die, but you will be preserved.

In much of the fiction which shapes our perceptions of sadistic cruelty the hero/victim's importance is a given. Only he or she can foil the villain's evil plans. The villain may seem obsessed with killing the hero, or may initially express derision, only to learn the error of his or her ways. In either case, the evil plans inevitably end in failure. If we must suffer, this is how we would like our suffering to be. Brief and endurable, it should also be part of a narrative in which we triumph in the end, having become immensely significant to our enemy en route. If we must be tortured, at least let us matter to the torturer.[28] The most popular fiction reassures us, just as religions do, with these three claims: that we have particular meaning and importance to others (even when they hate us); that cruelty is punished (except perhaps when we are the perpetrators); and that evil people are alien to us (and also to whichever powers ensure that justice will eventually be done).

Many of us are content to have our closest encounters with sadistic cruelty in the imagination. Yet some people, such as sexually motivated serial killers, choose to act out their savage fantasies. Sadism may be rarer than callousness, but it nonetheless exists. Worse, certain situations seem to encourage sadistic cruelty. We have seen that vicarious cruelty can be enjoyable, reinforcing our basic moral instincts and reassuring us that we are special. Are there other reasons why we like it—and why some people take their fantasies into the real world? To answer these questions we need to make the same transition, setting aside pretend cruelty for the real kind, and ask how sadistic behaviour could come about.

Sadism emerges from callousness

Sadism—unlike, say, reactive aggression—does not appear to be the kind of behaviour which human beings instinctively express *de novo* from an early age. Instead, sadistic behaviour builds on other forms of human harm-doing involving extreme otherization and callous violence. War, terrorism, gang violence, serial killing, slavery, and domestic abuse are examples in which sadistic cruelty arises out of callous cruelty. This is consistent with a typical pattern found in sadistic behaviour: gradual and often intermittent escalation from less severe acts of torment (such as beatings or acts against animals) to increasingly horrific acts of torture and murder. Like the otherization of a group, sadistic cruelty to individuals is most effectively prevented while still in its early stages.

Children's exploratory cruelty to animals also fits this pattern. Curiosity may motivate the initial behaviour, but it is the intervention of older, authoritative carers which provides the inhibition, demonstrating the moral rules which link actions to (in this case negative) consequences. As the child becomes socialized, those rules are increasingly accepted, treated as part of the self: one becomes a good citizen, a law-abiding person. This is a long, slow process, and it is far from straightforward. The urge to accept the rules of the community, and thus become more like other people, conflicts with the urge to challenge the rules, and thus become more different and independent.

'By a dichotomy familiar to us all,' a doctor remarks in John Wyndham's *The Midwich Cuckoos*, 'a woman requires her own baby to be perfectly normal, and at the same time superior to all other babies.'[29] Whether or not the conflict begins with mother love, the tug between sameness and difference, between accepting other people's beliefs and codes of conduct as yours and defining your own, can continue long after the second decade in which it typically reaches its turbulent heights. Teenage cultures emphasize the paradox. Outsiders see unwitting clones crushed under fashion's iron heel in the name of corporate profit: all those angry young people being resolutely *different* in tediously similar ways. Teenagers see a fierce struggle for freedom, identity, and respect. But all of us who share the symbolic world experience some form, however attenuated, of the social pressures which can make teenage existence such raw hell, and all of us constantly readjust ourselves accordingly. We examine new symbols, hypotheses, ideologies, assertions, and other influences, and we accept them as part of us, or not. We also use world-shaping, when changing the external 'that' or 'them' is easier than changing 'us'. And sometimes, in so doing, we do harm.

SADISM REQUIRES CALLOUSNESS TO BE LEGITIMIZED

Even for one sadistic individual the degree of cruelty shown can be highly variable. The German serial killer Peter Kürten let some of his victims escape, yet mercilessly butchered others. He thought about cruelty far more often than he acted out the thoughts, and he was able to control his urges when the situation warranted discretion.[30] As this intermittent eruption from fantasy into reality suggests, the nature of his environment—both at the time of the attacks and during Kürten's youth, was crucial in the expression (or not) of his sadism. Even for those of us who are not sexual sadists, certain situations, like war and slavery, reliably produce sadistic atrocities. Others, like cricket matches or academic conferences, generally do not. What is it about the former which contributes to the development of sadism?

The most obvious factor is a background of callous cruelty and frequent violence, a culture in which non-violent options are disregarded. Importantly, this high level of harm-doing is not seen by its practitioners as cruelty, in the sense defined in this book. It is considered legitimate, approved by their social role-models, and thereby socially acceptable. (Kürten, for instance, had a violently abusive father and claimed to have learned additional cruelty from his first employer, a dog-handler who liked to torment his animals.[31]) Callous cruelty is likely to be rewarded. It has become internalized as part of the perpetrators' self, their warrior toughness or firm authority. They may be soldiers who see themselves as doing society's dirty work, or parents who perceive their behaviour as enforcing discipline within the home. Whatever their role, they view pragmatically what others see as viciously unnecessary.

CALLOUSNESS BECOMES TERROR WHEN SUFFERING BECOMES USEFUL

Callous cruelty, to repeat, involves inflicting suffering in pursuit of other goals, and requires some degree of otherization: the devaluing of the victim's experience and human worth. Moving from callous to sadistic cruelty involves two further steps. The first is when causing suffering ceases to be merely a *consequence* of the perpetrator's attempt to achieve his or her ambitions, and becomes a useful *means* of achieving them. When suffering becomes instrumentalized in this way, perpetrators arrive at the rationale of terrorism: inflicting suffering on some people in order to punish, threaten, or deter both victims and, often more significantly, onlookers. This is not yet sadism, but the goal of inflicting suffering has become much more important, making the emergence of sadism more likely.

The cruelty of terrorism serves as a signal of power and intent.[32] It implies three messages, which may or may not be true. The first states: 'we are more powerful than you think' (since we can spare the time and energy to engage in this spectacular, bizarre, and inefficient means of harm-doing). The second states: 'we feel we have no alternative' (if we rationally choose to do something as vile as this, we must have extremely powerful reasons for doing so). The third states: 'you cannot persuade us to change' (so you must change to accommodate our demands). Terrorists operate on the assumption that both they and their enemies are capable of rational behaviour. They are not, on the whole, psychopathic, although a few may be sadistic psychopaths. Nor are they psychotic—or if they are they don't last long, as staging atrocities doesn't mix well with hallucinations and delusions.

This claim of sanity includes morality. Most terrorists acknowledge moral rules, and indeed rely on them both to communicate with their enemies and to garner support. The language of al-Qa'eda, to take a current example, is not like the language of a scientific report: value-neutral, with moral terminology minimized. Their talk is firmly located within the domain of right and wrong, good and evil, power and humiliation. The moral axis may be reversed compared to that of George W. Bush and his allies in the 'war on terror', but as numerous commentators have pointed out, the two are more like mirror images than we would expect, were al-Qa'eda's discourse *non*-moral.[33] (Some of the Islamic radicals' beliefs may seem delusional, but any criticism on those grounds should be spread evenly. We in the West are not above having—or acting on—delusions.)

Some Islamist demands would probably sound quite reasonable (e.g. 'stop humiliating Muslims') if only they weren't made by Islamists. These are people and organizations who avow, as the al-Qa'eda manual does, that 'peaceful solutions and cooperative councils' are useless against 'apostate regimes' which only understand 'the dialogue of bullets, the ideals of assassination, bombing, and destruction, and the diplomacy of the cannon and machine gun'.[34] This is the power and irony of otherization: al-Qa'eda's view of the West is pretty much how the West sees al-Qa'eda. Robert Burns' wish—'O wad some Pow'r the giftie gie us | *To see oursels as others see us!*'—is thereby granted.[35] Handed the mirror, we see our enemy: Us perceived as Them. The image shatters only when one side turns away from otherization, rejecting the stereotypes of demonic evil which justify atrocious cruelty.

TERROR BECOMES SADISM WHEN SUFFERING REWARDS
THE PERPETRATOR

The second step, which follows all too naturally from the use of cruelty as terror, is that the victim's suffering itself becomes rewarding. Somehow the perpetrator comes to find pleasure in the agonies which before were merely useful, and which for most of us would be extremely distressing. Eliciting those agonies becomes a priority, irrespective of other goals. Hurting victims has moved from side-effect to useful prop to centre stage.

Sadists can be useful to their superiors. They willingly do the jobs which might leave other personnel with post-traumatic stress disorder, enhancing the organization's ability to terrorize its foes. However, their focus on cruelty can also hamper their efficiency, dilute their commitment to other goals, and disgust their colleagues. Displaying sadism is thus a high-risk strategy. Applauded at one moment, or in private, it may bring down public wrath and rapid disavowal should circumstances change.

The need for control, like the empathy which reins it in, varies from person to person. This variability is normal; neither control freaks nor people with very low empathy are automatically destined for a life of violent crime. For them to become sadistic perpetrators, they must first become accustomed to callous cruelty—and not just accustomed, but accepting of callousness as legitimate behaviour. Secondly, their environment must offer little or no practical deterrent to their sadistic behaviour, either because those around them are similarly inclined or because the environment is so chaotic that moral rules are not enforced by punishment (or themselves become regarded as punishable weakness, as happens in many situations leading up to war crimes). Finally, the environment must specifically associate the actions which cause suffering with rewards—either through the individual's direct experience or through the observation that others who act in this way are well rewarded.

And well rewarded they certainly seem to be. Sadistic paedophiles, for instance, may delude themselves into justifying their behaviour, but they do so knowingly; it takes hard work to maintain those false beliefs. They are aware that other people will not share their carefully crafted moral judgements, and they world-shape, as we all do, in order to defend their precious justifications. This deliberate defence of cruelty by many sadistic perpetrators, let alone the effort involved in planning it, suggests that the rewards must be intense.

Here, however, we must be careful, because there are different kinds of rewards—and, confusingly, they can differ in various ways. The next step, therefore, is to look a little more closely at the pleasures which motivate cruelty.

The pleasures of sadism

> Evil, be thou my good . . .
>
> (John Milton, *Paradise Lost*)

Rewards differ in how they make us feel. The short-lived pleasure of tasting one's favourite drink is very different from the rush of cocaine, the thrill of a sky-dive, the pleasure of an unexpected cheque in the post, or the sweet contentment of a mother's hug. Some rewards are very intense, but tend not to last long (these are the ones which tend to provoke addiction); others linger, but aren't so overwhelming. Some are visceral, others more cerebral. Genetic differences and varying personal experiences lead human beings to differ considerably in their preferences for particular kinds of reward. Some people find delight in reading a book or doing something creative, while others prefer to be active; some value food and drink; some get their kicks from sex or friendship; and some enjoy the exercise of power.

Rewards also differ in their timescale. Some are immediate, some visible enough to be aimed at, and some completely imperceptible to the individuals who perform the rewarded action. These latter are the evolutionary benefits which, over time, lead to natural selection for that particular action, because on the whole, in general, people who act in that way tend to leave more copies of their genes than people who don't. The logic of evolution operates whether or not those engaged in spreading their genes can figure it out in any particular circumstance. As we saw earlier, the motivational gap (why do something if it brings no discernible pleasure?) tends to be filled by an intrinsic reward, such as sexual arousal or the sense of control: an enjoyable sensation which humans can link to actions whose benefits they may not have registered. Nature's deep reason why sex feels good is that it helps to spread genes, but people who practise recreational sex just enjoy the sensations. Genes are often said to be indifferent to us, but we can be equally indifferent to them.

These intrinsic rewards are biological pleasures: natural neural currency, supplied at birth. They can be activated directly, for example by intercourse, drug-taking, or fluent physical activity; though as the case of drugs shows, not all stimuli need be 'natural'. However, the brain's ability to form associations allows intrinsic rewards to exert influence far beyond their original domain. Other stimuli which may in themselves have no capacity to produce intrinsic rewards can become associated with them, simply by virtue of occurring around the same time, and can even come to evoke them entirely unaided.

This process of 'conditioning' was famously exemplified by Ivan Pavlov's dogs, who learnt to salivate on hearing a bell because the bell had previously been rung when food was imminent.[36] Bells and their noises are not edible, but the dogs associated them with edible rewards, and eventually came to behave as if the reward would appear even when it didn't. Humans can similarly learn to associate even disembodied symbols with powerful positive or negative feelings. A religious believer's perception of her deity, for instance, is likely to have much to do with the context and carers who taught her the faith (kindly parents make for kindly gods).[37] Her conditioning, especially if it takes place early in life, ensures that when she thinks of God the feelings she took from her carers return to claim her, even long after her teachers themselves are gone. An intrinsic reward, in this case the sense of secure belonging, has become conditioned to her concept of God.

We thus have three kinds of reward which could be relevant to sadism. Evolutionary rewards are benefits to genetic fitness (which apply to genes and may or may not affect individuals). Intrinsic rewards are directly activated brain pleasure systems (like orgasm during intercourse). Proxy rewards are intrinsic rewards conditioned to other stimuli, ideas, or behaviours (like orgasm during the act of killing, reported by some serial killers).[38] Which of these could contribute to sadistic cruelty?

SADISM AND SELECTION

As noted elsewhere, callous cruelty to outgroup members may indeed bring direct fitness benefits. Does sadism add any further value? One way in which it might draws on the second pillar in Darwin's mighty framework: sexual selection. Daniel Dennett calls such conjectures bowerbird theories, after a species whose males build extravagant nests, filled with interesting objects, to attract their mates.[39] Why? Because when one sex invests more in parenting it makes sense for members of that sex to choose mates carefully, weeding out losers. This in turn puts a selection pressure on the opposite sex to enhance those traits and behaviours which their fastidious partners find attractive.

Hence the extravagant mating displays of many males, which emphasize their physical health and/or the resources they have available, often at considerable risk to their safety. Anyone who has all but tripped over a male blackbird intent on seeing off a rival will recognize the utter recklessness brought on by the onset of the mating season. Females seem to seek out evidence of strength, good health, and physical symmetry, indicative of healthy genes, as well as signs of good resources and the behaviours expected of a father. Males who provide these may not always be rewarded, of course. I recently saw a pigeon struggle to heft a large twig towards his beloved ('Look! Nest-maker!'); she promptly flew off leaving him disconsolate, twig and all; perhaps she was holding out for a bowerbird. But males with preferred traits only need to be rewarded slightly more often than their rivals for evolution to take hold.

How might behaving cruelly raise men's chances in the sexual lottery? Sexual selection gives us two hypotheses. The first is that cruelty may impress females, causing them to mate preferentially with visibly vicious males. This 'James Bond' effect makes some sense in the case of callous behaviour, since females could benefit from favouring strong protectors who don't mind frightening off the competition.[40]

Gratuitously torturing potential suitors, however, might be expected to deter the ladies. Then again, that may depend on the ladies—which brings us to a sticking-point, because simple sex selection fails to explain active physical cruelty *by* females. Peahens may admire their partners' tailfeathers, but they feel no need to grow their own. We might expect women to approve of displays of sadism, but why would they partici-pate? Yet they do. Many notoriously cruel tyrants, and indeed multiple murderers, have found partners who condoned or joined in their sadistic behaviour. There have been female murderers (like Las Poquianchis, the Mexican sisters whose trial in 1964 convicted them of killing over ninety people).[41] Women have been suicide bombers, sadistic prison guards, even child abusers. Historically, far fewer women than men have committed acts of physically sadistic cruelty (they specialize in verbal, social, and emotional abuse). What is less clear is how much of the gap is down to gender differences, and how much of it due to social mores, evolved and acquired through culture, which punish and abhor physical (though not necessarily social) cruelty in women much more than its equivalent in men.[42] What is clear is that women can enjoy cruelty.

The second hypothesis is that cruelty by males may impress other males in a way that home-making skills can fail to achieve. Here sadism makes considerable sense, because it is a way of displaying excessive power. Not only am I able to kill my rival, but I am so secure and powerful that I can afford to take the time to flay him alive or pull out his fingernails first. Attack me if you dare, and risk the same fate. Displaying sadistic cruelty sends a message to potential challengers that you are unpredictable (you act on a whim, not from necessity) and powerful (you have enough excess energy to indulge in inflationary cruelty, so any attack will be met with disproportionate violence). You also come across as extremely dangerous, since attackers can't rely on your moral inhibitions kicking in during a fight to stop you from inflicting gratuitous injury or death.

If this hypothesis is correct, we should expect to see sadism reliably emerging in certain conditions: put bluntly, those in which male posturing is commonplace and callousness expected.[43] Some serial, mass, and spree killings may fit this pattern of inter-male competition, particularly when perpetrators consciously aim to surpass the excesses of their predecessors in order to achieve particular notoriety.[44] The escalation in suicide bombings observed when rival terrorist factions compete for authority in Palestine is another example.[45] Sexual selection makes sense in this context. It is not so much selection for sadism as for the ability to compartmentalize, torturing only some people and thereby protecting and benefiting others. Males who could be tender and caring to the ingroup while being viciously destructive to others would have more to offer a vulnerable female than either a domestic sadist or a well-meaning wimp.

CAN PAIN FEEL GOOD?

What of the second option: intrinsic reward? Can observing (or experiencing) pain itself directly activate the brain's pleasure systems? On this view, some acts of sadism could be committed by highly empathic and masochistic perpetrators. My guess, however, would be that this is not the general case. The aversive quality of pain is extremely deep-rooted, and it is hard to see what mechanisms could reverse that and turn pain into pleasure. Rather, it may be that other intrinsic rewards become associated with inflicting or experiencing pain (like excitement or sexual arousal), because they occur at the same time—and it is these which make sadism and masochism enjoyable, not the pain itself.[46] Killers and torturers may enjoy their victims' pain, but there is no evidence that they enjoy their own; moreover, reports from perpetrators suggest that they tend to suppress empathic reactions rather than trying to enhance them.

Research supports this: many violent criminals are poor at the emotional empathy needed to experience a victim's distress as painful, although some psychopaths, for instance, are skilful users of the cognitive techniques of theory of mind.[47]

Cruelty and control

The core of sadism, common to all its manifestations, is *the passion to have absolute and unrestricted control over a living being.*

(Erich Fromm, *The Anatomy of Human Destructiveness*)

That leaves us with the third option: proxy (conditioned) reward. Perhaps callous cruelty is like the desire for sugar: an excellent survival strategy for our ancestors, but problematic in the modern world. If being callous, like eating sugary fruits, provided a survival edge (fewer competitors or more energy, respectively), then humans could have evolved a desire analogous to their liking for sugar: a sensation which rewards them for being callous. An obvious candidate is the human sense of control: the feeling of exerting power over one's environment—including one's social environment. Like the urge to scoff sweets, the need for control varies with personality, gender, age, and cultural norms. It takes many different forms, and the rewards it offers can be intense, even addictive.[48]

The need for control is not the only intrinsic reward which can become associated with a victim's suffering. Sexual arousal can also become conditioned to distress signals, linking sex to physical violence. Cruelty which involves physical exertion, such as beating or chasing the victim, can bring the enjoyment which goes with physical movement and the excitement which goes with strenuous action.[49] The sense of control can also be boosted by the feeling of acting in unison with a group of like-minded others. Mirrored movements activate our motor systems, reinforcing the sense of having the power to act. They enhance group cohesion by increasing motor empathy. The endorphins and adrenaline we hold responsible for these sensations have no moral judgement; they make us feel good whatever the meaning of the actions we perform. In addition, a victim's distress may be interpreted as gratifying submission, particularly when empathy is weak or absent (and distress signals therefore less aversive), rewarding the perpetrator's need for control. Indeed, distress signals evolved to provide precisely this appeasement of affronted superiors.

Cruel behaviour may also bring social rewards. The urge to please group members, especially those more powerful than yourself, is an understandably powerful desire which has contributed far more to the grisly total of human evil-doing than sadism ever has. Conformity can provide the opportunity for sadism to emerge. A gang of children egging each other on to torment a puppy provides powerful positive reinforcement to group members and painful punishment to those reluctant to join in. The cheering citizens of Kovno, who laughed and clapped as Jews were murdered in front of them, no doubt exerted similar effects on those who delivered the victims to their deaths.

Intense positive emotions, particularly in highly stressful situations, can deliver potent rewards to perpetrators. Humour, which we count among the most distinctively human of our attributes, is also one of the commonest rewards of cruelty. Laughter is an effective social bond, uniting perpetrators and those who laugh along with them while locking out the already suffering victim. It is also yet another signal of power. In movies, for example, we admire and are meant to admire the wisecracking villain, despite our awareness of his crimes.

Sex and power are the motives that spring to mind when we think of sadism. Yet love and the need to belong can also fuel atrocious cruelty, whether by abused and unloved individuals or by members of the group who want to keep being liked. In principle, any intrinsic reward can become a reason for pursuing cruelty, even to the point of addiction. If sadism remains relatively uncommon, that is due to the extensive mechanisms developed in every culture to limit harm-doing and regulate rewards. Callous cruelty, however, is often exempted from these limits in threatening times, as basic morality overwhelms the constructed version. And when callousness provides intrinsic rewards there is always a risk that they will become more valuable to perpetrators than the goals which originally prompted the cruel behaviour. In practice, the boundary between callousness and sadism is not a firm one, given how fluid and opaque our motives can be.

We may think of the rewards of cruelty as exotic, malignant, and filthily carnal, the sweaty lusts of a sadist or a paedophile; but for callousness they need not be nearly so extreme. Everyday humour, social bonding, or boredom-driven thrill-seeking can suffice—but only in certain situations. As noted in Chapter 4, every potential action must overcome an inhibitory threshold in order to be realized. Individual adults immersed in the relative passivity of a civilian lifestyle generally need considerable prompting to push them over the threshold of even minor physical cruelty. Those who go on to become sadistic usually have pre-existing vulnerabilities, such as detectable brain dysfunction, which in effect lowers the threshold for cruel behaviour.[50]

Situations which give rise to sadism, such as war, do so both by rewarding callous cruelty and by lowering the inhibitory threshold further. Cruelty is made permissible by example, legitimate by otherization, and routine by repetition. The structural factors which allow this to occur are found in every war as well as in gang violence, revolution, and many other scenarios. One implication of this is that we should expect sadistic atrocities even in the wars our soldiers fight—however well trained in human rights those soldiers are.

Atrocities which seem entirely sadistic, of course, may actually be driven by other motives as well, especially when symbolic threats are involved and conditioning gives symbols the power to reward or punish. Just as stimuli signalling the victim's pain and distress—screams, pleading, blood, fear behaviour, and so on—can become enjoyable through association with intrinsic rewards, so can objects, images, even ideas. A kitchen knife is a useful way to slice onions; a kitchen knife interpreted as a weapon carries a much greater emotional weight. The same object, hefted by the same hand, can confer quite different feelings of control.

The need for control is also relevant to a commonly cited motive for cruel behaviour: the urge to transgress. Transgression involves breaking moral rules which have not been fully internalized by the transgressors (who are often immature, socially isolated, or both). The rules' symbolic power as important or even sacred codes is recognized; but the rules themselves are seen as somebody else's, imposed and resented. Breaking other people's rules devalues their power and makes the transgressor look and feel more powerful, giving transgression its excitement.[51]

Consider the following description of cruelty's effects. The victim was an older man, helpless against a group of attackers. He was a civilian, a father, socially respectable, and a member of a vulnerable minority. Worthy of respect, or at least toleration, one might have thought. This is his son's report of what was done to him:

> Then I saw an object sticking through the door and something that looked like a horseshoe. I walked over to the door and pushed it open. I saw that two horseshoes were nailed to two feet, and my eye followed the feet to the ankles, which were covered in blood, and then to the knees which looked disjointed. I looked up to the genitals, which were just a mound of blood, above which long snake-like lacerations rose up the abdomen to the chest. The hands were nailed horizontally on a board, which was meant to resemble a cross. The hands were clenched like claws around big spikes of iron driven into the board. The shoulders were remarkably clean and white, and the throat had a fringe of beard along the last inch of the body. There was nothing else on the cross. They had left the head near the steps to our house, just at the edge of the street. I could see his nose propped on the step. I could see the beard trimmed neatly along the cheekbones. I could see it was my father.[52]

Decades later, with no access to the perpetrators, there is much we can never know about the motivation behind this atrocity. Nevertheless, we may be able to see how an explanation could be possible—to see, in other words, that this cruelty, though vile, is not incomprehensible. The source is an Armenian memoir of events in 1915. The perpetrators were Turkish nationalists who wished to 'purify' their country of minorities, among whom the Christian Armenians were especially noticeable. The victims were highly otherized, presented in the Young Turks' propaganda as a classic disgust-threat, a festering menace undermining the nation's integrity. In such circumstances, as with the earlier example of the SS man and the baby, we should expect to see highly symbolic forms of cruelty, with evidence of world-shaping intended to transform a civilized and dignified victim into a physically disgusting object. The abdominal lacerations, bloody groin, and decapitation undoubtedly achieve that effect; and with the man's head removed, the most obvious features of his human identity have gone.

As for the symbolism, the Armenian Christian victim has undergone a parody of crucifixion. His genitals, source of male power and reproductive potential, have been rendered unidentifiable. This is transgression at its most brutal and obscene.[53] What we cannot know, without knowing the perpetrators' thoughts on the matter at the time, is whether their cruelty was callousness or sadism: the need to transform a challenging victim, or sheer predatory delight in power, blood, and pain. It may well have been a mix of the two. Does it matter? We react differently to cruelty depending on how we interpret perpetrators' motives—but should we? Callousness can be just as appalling for the victim.

Summary and conclusions

Humans developed morality and self-control to rein in the excessive aggression which led to callousness, but only against those they saw as human—similar to them and therefore likely to be kin. Callousness which targeted outsiders could bring not only long-term benefits for group survival, but more immediate rewards—new territory, food sources, or mates. These reinforced the callous behaviour and so sharpened the distinction between 'us' and 'them', developing the otherizing mechanisms ubiquitous today. In addition, callousness can have another pay-off: terror. Being cruel may waste more energy, but when the opponents are far less powerful than their attackers that hardly matters. The costs are outweighed by the deterrent effect, as potentially hostile third parties revise their perception of the attackers' capabilities sharply upwards.

The downside of such behaviour is familiar in the modern world. Survivors of callous cruelty often resort to extreme measures themselves, for revenge, for self-protection, or because they consider their attacker's cruelty evidence of less-than-human status. The resulting vicious circle, in which outrage breeds outrage and atrocity provokes not surrender but more atrocity, can create the ideal environment for sadism to flourish.

Sadism can take root when the distress signals which would normally inhibit aggression—such as tears, prostration, and signs of helplessness and humiliation—become associated with some form of reward, prompting the individual to enact behaviours which increase others' pain and distress. The reward can be so powerful that it promotes addiction. Like other addictions, however, sadism is triggered by specific environments. Just as drug addicts have a better chance of escaping their habit if they leave their old haunts for somewhere which doesn't constantly remind them of taking drugs, so sadistic behaviour during a war—or a relationship—may not occur outside those specific situations. Innumerable perpetrators have led the lives of good citizens after a war in which they committed atrocities.

This emphasis on situation is reminiscent of the arguments proposed by social psychologist Philip Zimbardo, whose Stanford Prison Experiment turned psychologically healthy young men into sadistic guards within a few days. Zimbardo interprets his notorious research as showing that doing evil is a temptation to which we are all susceptible, in suitable circumstances: 'we can learn to become good or evil regardless of our genetic inheritance, personality, or family legacy.'[54] Nevertheless, he also notes the individual differences in his guards' behaviour. Not all were sadistic. People do resist the temptation to be cruel. Not many show the outstanding heroism of individuals like Dietrich Bonhöffer, the Lutheran theologian who openly criticized Nazi anti-Semitism (and was hanged by the Nazis in 1945), but there were Germans who tried to avoid complicity with the regime, whether that meant deliberately misidentifying Jewish children, smuggling out reports of atrocities, emigrating, or simply keeping one's head down, not getting involved in the denunciation and persecution of local Jews.

Even among SS men, the Nazi military elite, there were differences. Those sent to the Eastern Front were confronted, sometimes without warning, with extreme situational pressure (being ordered to kill Jews by their superior officers). They showed a variety of reactions. Some SS men came to enjoy the killing, although most saw it as a more or less hideous chore, to be endured with the help of alcohol and/or ideological support. Some, however, detested it so much that they refused to participate and were transferred to other duties.[55]

No single atrocity can be fully explained in scientific terms by the theories currently available. We simply do not have the data. But that does not mean that atrocity in general, as a phenomenon studied like any other, is inexplicable. We can identify common features and potential mechanisms, such as world-shaping, disgust-threat responses, and the processes of otherization, which lead to cruel behaviour. We can also begin to see that human evil, even in its most hideous manifestations, is not incomprehensible. That is alarming, because it brings the capacity for cruelty close to home.

Understanding cruelty, however, is the only viable means, in the long term, of dealing with it. Too often the lack of understanding itself contributes to the otherization which pushes groups apart and whips up the motives and reasons for acting atrociously. Recognizing the warning signs and patterns of cruel behaviour and acknowledging our own susceptibilities may help us put in place the beliefs and the social incentives which can make cruelty less likely to occur—though it will never be eliminated. If we can come to see cruelty not as otherized evil, but as comprehensible, if morally vile, behaviour, it may be a little less painful to grasp that being cruel is part of being human.

Chapter 9

Can we stop being cruel?

Not only had we become criminals, we had become a ferocious species in a barbarous world. This truth is not believable
to someone who has not lived it in his muscles. Our daily life
was unnatural and bloody, and that suited us.

(Pio Mutungirehe, a perpetrator in the Rwandan genocide,
cited in Jean Hatzfeld's *A Time for Machetes*)

A science of cruelty

I have argued that applying findings from the brain sciences to the study
of cruelty is both feasible and useful. Using neuroscience and psychology
allows us to think about cruelty's components (e.g. beliefs, emotions,
and actions) and mechanisms (e.g. otherization and threat responses) and
to constrain our ideas about them using our growing understanding of
how brains work. We can relate cruel behaviour to well-established theoretical frameworks like evolution and the synaptic learning paradigm of
neuroscience. We can also bring together existing knowledge and methods from fields as disparate as neuroanatomy, evolutionary psychology,
cognitive neuroscience, and social psychology to be combined in truly
interdisciplinary models. These increase our understanding of one of
the most difficult human problems, helping us to identify the causes of
cruelty.

In Chapters 4, 5, and 6, for example, we explored a view of brain function which sheds new light on how human beings can come to act cruelly. That view still sees people as acting for reasons which seem good to them, but the computer-like view of rationality as logical, impassive, and fully possessed of relevant information has been replaced by a rather different form of reasoning. Partial, intuitive, ignorant, passionate, and hasty, our choices are swayed by short-term advantages, immediate desires, situational pressures, a tendency to forget that other people's lives and experiences may be as rich as ours, and a very limited ability to consider the consequences of our actions.

This is not to say that logical thinking implies higher moral status. Far from it. Those decision-makers, safe and leisured in their offices, whose choices lead to others being cruel may be able to take more logical approaches to their problems, but being logical about humankind is entirely compatible with being callously cruel to particular humans—especially humans in large numbers. If the history of the twentieth century has taught us anything it is that reason and kindness are not always linked, much though we might like them to be. It is the unreasonable fallibility of a perpetrator's brain which allows him to be brought up short by sudden empathy.

That same fallibility, however, leaves all of us open to being cruel at times. The way our brains work means that stimuli—even the terrible stimuli of human suffering—can have decreasing impact on us simply by happening again and again. Likewise, repeated actions (especially those aimed at relieving strong negative emotions) become easier to do and harder not to do. Violent responses can thus become ingrained. Since even imagining cruel behaviour activates the neural patterns involved in acting cruelly (albeit less strongly than the actual behaviour), thinking or talking about cruelty can make it more likely—if the thinking or talking is rewarded. If it is punished, of course, then the behaviour can be made less likely to occur. But if the environment (physical or social) approves a person's otherizing actions, then repeated exposure to otherizing stimuli—even if it does not lead to cruel behaviour at the time—can strengthen the relevant patterns. This makes future stimuli more likely to trigger explosive violence, sometimes in response to apparently trivial provocation.

We also learned, in Chapter 6, that the role of strong beliefs in cruelty has much to do with how brains are constructed. Bolstered by their connections with other neural patterns and/or inputs from the body, patterns which ground some beliefs can grow strong enough to escape the influence of inconvenient truth. Not only can evidence and reasoned argument come to seem irrelevant, but the beliefs themselves can feel more certain, and the objects whose existence they assert more real, than reality itself. Dismissing such beliefs as deluded, challenging them directly, or promoting conflicting beliefs may be satisfying to the challenger, but will probably strengthen rather than erode the beliefs themselves—especially if the challenger is already considered an enemy by the believer. Criticizing strong beliefs may thus backfire by stimulating both a search for justification and the vicious spiral of otherization.

Like any other neural pattern, the patterns grounding strong beliefs are constantly adjusting their synapses to reflect their levels of activity. This process involves a balance between two factors: the current situation and the pattern's accumulated history. The stronger that pattern, the heavier its history weighs in the balance. In a world before symbols, that history would have been amassed either by long repetition or by the sudden shock of serious threats. In either case it made sense to trust the past experience which had, after all, resulted in continuing survival. Today, however, strong beliefs can be created without the need for lengthy prior experience or actual threat. Indeed, they can become so strong that even contradictions can be absorbed without surrendering the dogma. Faced with inputs which would force less committed others to adjust or abandon their ideas, a strong believer may find it less painful to adjust (or sometimes abandon) that bit of the world which gave rise to the offending inputs. World-shaping may lead to abhorrent cruelty, self-protective for the perpetrators, incomprehensible to those who do not understand or share their point of view.

Cruelty, like any human behaviour, is mediated by brains. The brain sciences, however, are not sufficient to understand it fully. Cruelty is also a moral concept, reliant on notions of justification and necessity, undeserved suffering, and voluntary, intentional action. The decision to call a person or action cruel, like other decisions, is inflected by various psychological and social biases. Thus judgements vary depending on who makes them; this kind of justice is very far from blind. Cruelty must therefore be set in its symbolic context with help from historians, anthropologists, philosophers, sociologists, and cultural theorists. Neuroscience and psychology, important though they are, are partners in a greater enterprise, whose aim is not to 'explain away' cruel behaviour but to learn enough to reduce the harm it does.

Cruelty can be thought of as dimensional, ranging between extremes of pure callousness and pure sadism, depending on the motives involved. Callous perpetrators disregard their victims' suffering, more or less effortfully (less, with practice). If the suffering is more valuable to the perpetrator—for example, because it is seen as punishing, deterring, or as sending a message about the perpetrator's power and determination—then his or her cruelty is correspondingly more focused on the production of suffering; this is the realm of inflationary cruelty and terrorism. At the rare extreme of sadism the victim's anguish ceases to be inflicted for other reasons than its own reward, and thus becomes the aim of the exercise.

Cruelty's rewards are various, because human physiology incorporates multiple systems for delivering intrinsic rewards. Food and other chemicals, sexual release, material gain, power, and even love can act as rewards to reinforce cruel behaviour. Cruelty can feel like just punishment for outrageous wrongdoers, and delight in pain like no more than a twinge of *Schadenfreude*; but not all of cruelty's pleasures present themselves as righteous. Perpetrators may feel excitingly transgressive, erotically aroused, or wondrously, blatantly powerful. They may feel a sense of belonging, even self-transcendence, immersed in the group. They may be reassured by the restoration of social order, relieved by the reduction of mental conflict, or wearied by the chore of what they do. Or they may feel nothing much at all.

Implications

The scientific study of cruelty has consequences for our everyday understanding of human harm-doing. If we can come to grasp why people commit atrocities, we may be able to prevent them—that is the hope. To see this, let us consider some of the most significant implications arising from the scientific approach.

CRUELTY ARISES FROM HUMAN FAILURE

We think of cruelty as centred on malice and hate, but it is more about failure. This is particularly clear when we look at organized, large-scale cruelty, whose perpetrators often appear to feel remarkably little hatred for their enemies. Failure haunts all such atrocities. Political leaders fail to challenge false beliefs which otherize their enemies, or fail to prevent illegal persecution of target groups. They may also fail to clarify their own intentions and desires, for fear of being held accountable for their actions. Often no explicit orders need be given for the leader's desires to be obvious to his followers. Individual perpetrators also fail. They do not check their beliefs against reality; they ask *cui bono* of their enemies but not of those they like and trust, and they succumb to implied threats of punishment without asking how likely or severe those punishments really are. The immediate rewards of going along with the group are ranked highly, while victims' pain is downplayed, denied, or defensively mocked. The final failure is that of bystanders. They do not teach their children to stamp down on cruelty before it can be expressed. They indulge mild otherization in themselves and others, either from social discomfort or because 'it's only a joke, what harm can it do?'. And when otherization spirals into large-scale violence they do not intervene, or sometimes even comment, as the atrocity is planned and perpetrated.[1]

People are often cruel because cruelty seems the easiest course of action in their circumstances. There may be very strong reasons for not being cruel, but at times it takes less effort and courage to ignore these than to resist the pressure, real or perceived, from other perpetrators. Indifference, laziness, wilful ignorance, and fear of the unknown consequences of lifting one's head above the parapet can be heightened by deliberate manipulation: for example, when leaders leave the consequences of speaking out unclear. The irony is that social pressure may itself be largely imaginary. Perpetrators unhappy with their role may keep silent rather than risk the group's united displeasure, but their silence itself inflates the perception of consensus. Individuals with social authority can make a huge difference by the values they choose to emphasize—but since that authority depends on their community accepting them as leaders, they may be afraid to risk group support by speaking out. When groups feel under threat (however realistic that threat may be), they emphasize values which strengthen their sense of control, like unity, loyalty, and sacrifice, rather than values which lessen group power by heightening autonomy and moral awareness.

If we are to reduce the prevalence of cruel behaviour, large-scale or small-scale, we will need to change the incentives for being cruel, so that such failures incur more negative and fewer positive consequences. Easily said, this is immensely difficult to do.[2] Yet it is not impossible. The English legal system once condemned people to public execution for stealing food. In the twentieth century, however, British culture rejected the spectacle of public execution even for much worse offences. Many in Britain would like to see capital punishment reinstated, but very few, I suspect, would want it used for minor theft. Thus far have we come.

Nonetheless, in both factual and fictional media, cruelty is still frequently associated with glamour and excitement rather than with its traumatic effects on victims. Easily available technology makes serious cruelty easier. Furthermore, poor social enforcement, whether by parents, the legal system, neighbours, or opinion-formers, undermines the ancient link between hurting and being hurt in return. We are all very scared of punishment nowadays, precisely because we recognize how easily it can escalate into cruelty, so we increasingly restrict its range to financial damage, community service, or loss of liberty. In terms of strengthening inhibitions within the brain, however, these options are somewhat problematic. They may not carry the required emotional weight, especially for experienced offenders; and their association with the offence itself may be very weak, in neural terms, because of the long gap between committing the offence and receiving the punishment (if indeed the offender is caught at all).[3]

No amount of rewarding people for good behaviour will necessarily stop them committing bad behaviour. To do that requires the inhibiting awareness that bad behaviour is liable to be punished. And to be effective the punishment must hurt. That pain could be the shocked recognition of a parent's disapproval or of a victim's anguish; it need not be physical. Yet without it, punishment succeeds only in pleasing those who punish. What this implies is that effective socialization early in life, as a method of crime reduction, is distinctly preferable to later punishment, since children are influenced by mildly aversive stimuli (such as parental disapproval) to which older individuals can be entirely impervious. Applying our punishments, be they only disapproval, when otherization is first apparent has more chance of success than waiting until it has triggered violence and then reacting. That applies whether the perpetrator is a schoolchild shouting abuse or a dictator applying disgust-language to his enemies. There is much we can do to stem cruelty, if we act promptly.

CRUELTY IS NATURAL BUT NOT IMMUTABLE

Human cruelty emerges from the way humans have evolved to survive as social animals. It is a product of the same capacities and motivations which have given us extraordinary control over our environments, allowing at least some of us to lengthen our life-spans, reduce the suffering caused by disease and malnutrition, dispel the threats which gave our ancestors nightmares, and invent a whole new set of anxieties. Cruelty is thus as natural as laziness or competitiveness. It protects the precious self, physical or symbolic, by doing harm to those about whom we care less. Basic morality, unlike the constructed 'human rights' variety, does not agree that all human beings are equal.

Some degree of cruelty is inevitable in human existence. All we can do is firefight—but as noted above, we could do far more than we are currently doing to minimize the suffering we cause each other. Unfortunate individuals may be driven by disease or damage or hideous experience to perpetrate spectacular sadism, but the vast majority of perpetrators are not so compelled. Most cruelty is callous, not sadistic (and many people learn to live with minimal cruelty to others). It is a behaviour—something people do, not something they are—and hence susceptible to incentives, the rewards and punishments which situations offer and groups impose on group members. For example, cruelty in response to a perceived attack on some cherished aspect of identity should be reduced if the threatened self can be bolstered by other methods, such as pondering its excellence in an unrelated domain.[4] The right manipulations, given the political will, could greatly lessen victims' suffering.

That the requisite determination is lacking has little to do with technical feasibility (although there is still much to learn). More relevant is the extent to which cruelty has been displaced to particular experts—those who grow our meat and fight our wars, for instance—and contained in cultural ghettos—'troublespots' like Israel/Palestine and Somalia or, more locally, 'problem neighbourhoods'. Otherization says cruelty happens there but not here; to them but not to us; is done by others but not ourselves. This allows us not only to abdicate responsibility but to pretend that cruelty does not exist, except of course on those rare occasions where it escapes its boundaries and claws its way into our lives. Add to this the enormous vested interests making money from human cruelty, from arms manufacturers to the aftermath industries which clean up after conflicts, and our tolerance of atrocities starts to make more sense.

OTHERIZATION CAN BE USEFUL

Otherization evolved because it was useful. It allowed groups to win competitions for scarce resources by bolstering their unity and power and shutting down their inhibitions about competitors. Otherization is still popular as a way of resolving conflicts; dead men cannot argue or retaliate. Our addiction to symbols, however, has greatly widened its range, allowing us to be manipulated into destroying threats which may not actually exist. The Nazi allegation of a Jewish-Bolshevik conspiracy was false, but hugely influential. It provided a target for anger and anxieties evoked by factors beyond individual control, offering simple explanations which tapped into long-standing prejudices and so were more easily accepted. It proposed actions, so that by acting—and especially by acting together—individuals could feel more powerful; and it enhanced their sense of control still further with bombastic rhetoric and paraphernalia. No wonder Adolf Hitler was cheered through the streets, at least in the early stages of his government.

Childhood socialization, if done adequately, teaches the deep and painful lesson that cruelty hurts the person who is cruel. Education adds a layer of moral principle, but the moral knowledge that cruelty is wrong is, for most of us, only a thin skin between thought and action, however much we might like to think otherwise. By accepting our own frailties and acknowledging that otherization can be useful, we can take the first steps towards defending ourselves against the temptation to use it. That defence is personal, but it is also social. Imagining cruelty, whether as an individual or in public debate, can make cruelty more likely to happen—if the cruelty is linked to the glow of rewards. Yet if it is not, open consideration of potential cruelty can make us aware of alternative paths of action. This destroys the necessity defence which perpetrators so frequently invoke. Imagination also makes space for cruelty to be condemned before it happens, and for people to see what interventions on their part could prevent it happening. Early intervention is preferable; it takes less effort.

CALLOUSNESS IS ENHANCED BY HUMAN BIASES

Human beings have evolved to treat the world unequally in several respects. We naturally care for close kin (or symbolic kin), for people we see as powerful ingroup members, and for beliefs which fit with ours. We naturally care less for strangers, low-status people, or ideas which contradict what we believe. We are limited information-processors, ignoring much of what our senses tell us because it does not suit us to believe it. We also bias our world-views over time, regarding distant future consequences as less interesting and relevant than current or upcoming events (a phenomenon psychologists call temporal discounting).[5] When we choose to be cruel, therefore, we often choose on the basis of very partial input, failing to consider many consequences of the harm we do. Cruelty's rewards are typically not long in coming: the booty is there to be snatched, the victim's perceived threat removed, the group's approval obvious. Cruelty's punishments usually take much longer, if they happen at all.

Being afflicted by biases, however, does not imply a fixed and disastrous nature. Biases vary from person to person, and that implies that we can learn to change them, at least to some extent. Again, knowledge is power; we have to recognize that biases exist before we can begin to work against them. Our brains are not set in stone, but are sensitively responsive to experience. Meeting the people we despise, facing new evidence, valuing open-mindedness rather than deriding it as 'wishy-washy'...once again, there is much we are already doing and much more we could do to overcome our basic moral instincts.

SEEING CRUELTY AS EVIL IS MISLEADING

Otherization leads us naturally to think of evil, for which sadistic cruelty is our most prominent marker, as essence rather than behaviour. In other people, evil can be viewed as a quality of their character or nature, unchangeable and repugnant. In oneself, or those one cares for, acknowledged evil is tolerable only as non-self, whether it is downplayed, denied, seen as 'possession' by some wicked entity, or attributed to vile necessity. If we were cruel, we were always driven to it. This is a dangerous justification of our cruelty to others we think cruel (how else can we remove their threat?). It encourages us to view our cruelty as some kind of optional extra—an alien infestation which in principle can be removed to leave us spotless.

If the Nazis had wiped out every Jew in Europe, would Hitler's problems have come to an immediate end? If every religious believer could be killed or made an atheist, would that solve the problems which still afflict our world today? Of course not. The problems would remain and new scapegoats would have to be found. Otherization is always self-defeating in the end, proving its falsity by its very success. Modern societies, with their extraordinary levels of cooperation, negotiation, and compromise, demonstrate by contrast the potential released when otherization is limited.

Yet ego and ignorance, lazy thinking and the inflamed need for control can lead people who seek to use otherization to disregard its built-in habit of failure. Much cruelty is about feeling powerful, and like all appetites the need for control is subject to temporal discounting, wanting satisfaction *now*, not in some long-term future. Power is also exciting, and displays of cruelty can be extremely exciting to onlookers; in that sense the glamour of cruelty is real, not mythical. Essentializing cruelty as evil has the perverse effect of enhancing the glamorous aura by presenting cruelty as an active, malevolent, and powerful force, not the outcome of lazy and limited human decision-making. Paradoxically, it also stops us taking cruelty seriously. We may see cruelty as evil, and thus as 'other', abdicating our responsibility to change our own behaviour. Or we may laugh at the primitive notions of our fellows and deny the moral framework of good and evil which gives the concept of cruelty its meaning, thereby becoming relativist sophisticates who blame human harm-doing on social inequality, poor parenting, anything except the people who actually do the harm.

Here again education can help to reduce the power of false beliefs. Learning facts and learning how to think can help us to see through the myriad forms of nonsense propounded by people who have much to gain from otherization. We can ask *cui bono?* We can also ask for evidence that the slurs against our enemies, or our minorities, are true. We can notice when claims depend on essentializing others, and reject them; we can challenge stereotypes. We can also teach our children better history. This is not just about the flagellating recital of past wrongs, whether committed by Britain, the United States, Germany, Japan, or others; every state likely to feature on the curriculum has perpetrated atrocities and suffered from them. Instead, learning history can set those crimes in context, showing that there are reasons why they happen and offering the hope that they may be avoided.

BELIEFS ARE CENTRAL TO CRUELTY

How do we learn what to value, what to desire? Some of our needs, like the urge to eat and the search for control, are intrinsic, but many more are acquired through learning what other people cherish, particularly people we like and respect. When we inflict cruelty in order to achieve desires or defend beliefs, those desires and beliefs are largely other people's interpretations, superimposed upon our own inchoate neural patterns. As we learn to identify and describe our thoughts and emotions, we change their patterns (the brain's version of the measurement problem in quantum mechanics, in which observation changes what is observed). Social learning—or post-hoc justification—helps us to see ourselves as others see us, but it also shapes us to better match their perceptions. To the extent that we adopt other people's expectations as our own, we become what they expect us to become, sharing their beliefs and wanting what we have been taught to want.

This docility is a vital component of modern existence: it keeps us social. But not all needs are as imperative as we have learnt to believe. Unfortunately, beliefs which are challenged by conflicts with other people can come to be more highly valued in consequence—sometimes even elevated to sacred principles. Some beliefs, furthermore, inevitably lead to conflict, such as the idea that we all have the right to have our strongest beliefs respected, whatever their content. People who profit from sacralizing our beliefs and desires (presenting them as unchallengeable 'truths' or 'rights') quite reasonably encourage such sacralization, promoting beliefs about the primacy of individual emotion and personal rights, equating critique and debate to disrespect, implying the usefulness of wealth and physical beauty as markers of moral virtue and social power, and assuming that growth and consumption are good in themselves. A skilled propagandist can make this sound like a coherent philosophy, though in most cases where the message is put across (e.g. advertising) no reasons are given for why we should believe it.

Does wealth make us happier, better people? Is a challenge to your favourite dogma always meant as a personal attack on you? Do our rights matter more than our responsibilities? These are debatable beliefs, not settled truths. To debate them, however, we first have to recognize their importance and their relevance to selfish and cruel behaviour. We also need the mental equipment to be capable of judging an argument's worth, and the will to challenge ideas and risk the discomfort of conflict, cognitive and social. Given all these, we can weaken many prevalent beliefs by challenging them, as some of our predecessors did by objecting to slavery, arguing for women's rights, or mocking the stupid beliefs underlying racism. Not all beliefs, as we have seen, can be dispersed by challenges. In many cases, however, a doctrine's apparent strength is largely illusory, inflated simply because it has rarely been queried. Once good arguments against it are put forward, a culture can ditch a belief with startling speed. Beliefs which encourage cruelty do not all show the mind-controlling fierceness of fanaticism. Quite often they are mere assumptions, weakly but widely held—and therefore vulnerable, if enough dissenting voices can make themselves heard.

CRUELTY, AND MORALITY, ARE ENHANCED BY EMOTIONS

Strong feelings, whether of terror, outrage, greed, disgust, or excitement, can fuel extreme otherization, turning punishment to vengeance, law to mob rule, and self-defence to ferocious cruelty. Ideally, such transitions to violence should be prevented before they occur, but strong emotions are notoriously difficult to defuse. Given a strong enough incentive (like the threat of death), most people can control themselves, and emotion regulation can be learned—if the person has the time, resources, and inclination to learn, or the power exists to make the threat believable. As otherization spirals towards cruelty, however, it creates pressures for more and more rapid action, destroying any chance to stop and think. It also makes participants feel more powerful, and thereby less afraid of the power of others.

One alternative is to counteract otherization with inhibitions which are themselves fired up with strong emotions. In neural terms, this means strengthening the inhibitory patterns whose activation tends to block cruel behaviour—the ones we associate with moral beliefs and with empathetic awareness of the victim's humanity. It is not enough to reassert an abstract moral principle; the power of emotions must be deployed. One way to do this is to link a person's moral beliefs to his or her sense of identity. People who resist the pressure to be cruel often seem to regard their belief that cruelty is wrong as part of their being and the demand to be cruel as a threat to their entire self. People who think of themselves as individuals—or as group members—who possess particular moral beliefs, on the other hand, can jettison those beliefs without risking such immense psychological damage, because the sources of their identities lie elsewhere.

If moral beliefs are made extremely strong, they will be activated whenever the prospect of cruelty confronts the person. If they remain at the level of abstract principles, they may not be activated at all (for example, because the person does not recognize their relevance to the situation), they may be activated too late, or they may be activated in time but not strongly enough to prevent the cruel behaviour from being triggered. This may lead to perpetrators feeling remorse, which may or may not assist their moral improvement. It cannot repair the harm inflicted on their victims, but it may deter them from future cruelty.

CRUELTY MAY BE SUSCEPTIBLE TO TECHNOLOGY

We have used information about how the brain works to consider tactics which may reduce cruel behaviour. A neuroscientific approach to cruelty, however, leaves open another possibility: that in some not-too-distant future we may be able to modify brains directly. Memory modification using chemicals has already been done in experimental animals.[6] There is no reason why precision removal of strong and dangerous beliefs in humans should not soon be a realistic treatment. Just imagine being able to cure the bigots in Israel and Palestine, damp down the Islamists and the racists, convert the homophobes and sexists to reasonable folk—all with a pill or a spray or some other quick fix, instead of the wearisome slog of changing beliefs and re-ordering incentives. Just imagine the ethics of trying to test such treatments in advance, and the delight with which your government would fund the research. Imagine, moreover, the haste with which your military would use it and the speed with which the definition of 'dangerous' would expand. Few governments can resist the temptation to see beliefs which contradict them as a problem. If your state obtained the power to edit belief, how long would your ideals be left untouched?

The capability to control cruelty by directly interfering with the brain is, to say the least, a double-edged sword. Interference can work well with objective facts, like brain tumours or meningitis, but cruelty and bigotry are not so tractable—because of their moral dimensions. A person who ignores these moral issues naively assumes that his or her moral judgement of which beliefs are good and which evil is the best available. Whether scientist, politician, or media commentator, that assumption is likely to be wrong; members of these professions are not known for their moral excellence. At the very least, we need a far more urgent public discussion than currently exists about the ethics of these identity-threatening technological advances.

Cruelty, science, and morality

A key argument of this book has been that, in understanding cruelty, morality matters. A science of cruelty by itself is not enough; we must also consider the moral and social aspects of human harm-doing. But is this the case? If every action can be sewn into a causal network, then do moral claims make any sense? Perhaps morality, wilting under the bright clear light of reason, should be discarded, like that other evolutionary relic, religion. This approach, which I once heard described as 'so very twentieth-century', can in fact be found much earlier: in the Enlightenment, expressed by a man hailed as 'the freest spirit who ever lived'.[7] So how does the Marquis de Sade attack morality? Let us consider his challenge in his own words.

SADE'S CHALLENGE

> Nature, red in tooth and claw
>
> (Tennyson, *In Memoriam*)

The nihilistic excesses of Sade's novel *Justine* contain philosophizing as well as pornography. Throughout the book this most determined of controversialists lays down a challenge to his society—and ours. Here it is in all its sardonic vigour:

> The man endowed with uncommon tastes [i.e. for cruelty] is sick; if you prefer, he is like a woman subject to hysterical vapors. Has the idea to punish such a person ever occurred to us? let us be equally fair when dealing with the man whose caprices startle us; perfectly like unto the ill man or the woman suffering from vapors, he is deserving of sympathy and not of blame; that is the moral apology for the persons whom we are discussing; a physical explanation will without doubt be found as easily, and when the study of anatomy reaches perfection they will without any trouble be able to demonstrate the relationship of the human constitution to the tastes which it affects. Ah, you pedants, hangmen, turnkeys, lawmakers, you shavepate rabble, what will you do when we have arrived there? what is to become of your laws, your ethics, your religion, your gallows, your Gods and your Heavens and your Hell when it shall be proven that such a flow of liquids, this variety of fibers, that degree of pungency in the blood or in the animal spirits are sufficient to make a man the object of your givings and your takings away?[8]

What indeed? If physical causes stir all our beliefs and desires, then we are surely just puppets of Nature, Sade's deity, who has made us 'isolated, envious, cruel and despotic; wishing to have everything and surrender nothing'. If so, selfishness, even to the point of murder, is surely natural. 'One must never appraise values save in terms of our own interests. The cessation of the victims' existences is as nothing compared to the continuation of ours...there is no rational commensuration between what affects us and what affects others; the first we sense physically, the other only touches us morally, and moral feelings are made to deceive; none but physical sensations are authentic.' Nature herself is indifferent to us, and we, if we are to be fully natural, fully 'authentic', should be equally indifferent to one another:

> ... every form is of equal worth in Nature's view; nothing is lost in the immense melting pot where variations are wrought: all the material masses which fall into it spring incessantly forth in other shapes, and whatsoever be our interventions in this process, not one of them, needless to say, outrages her, not one is capable of offending her. Our depredations revive her power; they stimulate her energy, but not one attenuates her; she is neither impeded nor thwarted by any.... Why! what difference does it make to her creative hand if this mass of flesh today wearing the conformation of a bipedal individual is reproduced tomorrow in the guise of a handful of centipedes? Dare one say that the construction of this two-legged animal costs her any more than that of an earthworm, and that she should take a greater interest in the one than in the other? If then the degree of attachment, or rather of indifference, is the same, what can it be to her if, by one man's sword, another man is trans-speciated into a fly or a blade of grass?[9]

Nature is a callous goddess. Faiths with more interested divinities have traditionally provided the dominant justification for not killing people, and Sade is well aware of this. 'The doctrine of brotherly love', he writes, 'is a fiction we owe to Christianity and not to Nature; the exponent of the Nazarene's cult, tormented, wretched and consequently in an enfeebled state which prompted him to cry out for tolerance and compassion, had no choice but to allege this fabulous relationship between one person and another'. Sade loathes Christianity, referring to it as 'that appalling cult' and describing religions in general as 'Mysteries which cause reason to shudder, dogmas which outrage Nature, grotesque ceremonies which simply inspire derision and disgust'. As for the concept of God, it too gets short shrift:

> Examine for one cold-blooded instant all the ridiculous and contradic-
> tory qualities wherewith the fabricators of this execrable chimera have
> been obliged to clothe him; verify for your own self how they contra-
> dict one another, annul one another, and you will recognize that this
> deific phantom, engendered by the fear of some and the ignorance of
> all, is nothing but a loathsome platitude which merits from us neither
> an instant of faith nor a minute's examination; a pitiable extravagance,
> disgusting to the mind, revolting to the heart, which ought never to
> have issued from the darkness save to plunge back into it, forever to
> be drowned.[10]

By now you may be thinking that the sentiment, if not the style, is so reminiscent of modern-day atheists that it may be time to believe in reincarnation. Resist the temptation, for here the two diverge. Not only are leading 'brights' not practical Sadeians, they have a much more positive view of Nature. They see faith in science and reason as justification enough for civilized morals, appealing to evolutionary psychology to argue that in fact nature made us nice—or at least, gave us the capacity for some limited degree of niceness.[11]

Sade takes a darker view. He sees the natural world as replete with cruelty, where animals and humans kill each other without compunction and often inflict prolonged agony in the process. His challenge, to anyone attempting to justify moral conduct, is to explain why we should take account of moral feelings at all. Why bother with the moral framework, with its restrictive prohibitions, its anxieties and guilts—even the group-centred kindness of basic morality, let alone its constructed variants? Why not simply rely on self-interested thinking, assessing costs and benefits and acting accordingly? Even if Sade is wrong about 'authenticity', and even if kindly feelings, as evolutionary psychologists argue, are part of human nature, that does not mean we have to take them as a guide to conduct. Vilely immoral behaviour appears to be just as evolved. Besides, having an appendix is part of human nature; but we get our appendix cut out as soon as it starts threatening to harm us. Maybe we should do the same with moral feelings, nice and nasty, and rely on self-interest instead—and if self-interest says 'Try cruelty!', so be it.

Sade's challenge contains two important strands. The first—'a physical explanation will without doubt be found as easily'—is none other than that ancient philosophical conundrum, the problem of free will. If everything we think, feel, and do is predetermined, part of a causal chain, how can we be the agents morality requires? The second strand—'every form is of equal worth in Nature's view'—asks why we should bother being moral. Religion gives us God(s) as the fount of moral authority, but once 'that execrable chimera' has been dispatched why should our morals have any authority at all?

THE PROBLEM OF FREE WILL

So much has been said about free will and determinism, by philosophers, scientists, and others, that I could not possibly review it here. Suffice it to say that some thinkers argue that free will is not required in order to be able to act morally.[12] If that is the case, your position on free will need have no impact on your view of cruelty.

And yet, and yet...We contemplate the wonders of modern brain science—whose eager press releases seem to hint that, given just a little more funding, scientists could figure out this ball of sludge and build a better one—and we see mystery after mystery snared in the nets of mechanistic explanation. It is in the brain, after all, that we seem best able to trace determinism's causal chains. If the 'bits in between' perception and behaviour are just a matter of causal 'constant conjunction', as David Hume argued, or statistical correlation (the modern version of Hume's revolutionary suggestion), then aren't selves just places where causes trigger their effects, rather than the agents we feel them to be?[13] If there are to be no such ontological certainties in our heads, then surely morality is only one of the comforts we must jettison. If selves are unreal, what is it that massacres defenceless victims? If inputs trigger outputs, can perpetrators be fairly held responsible for the atrocities they commit?

First of all, let us scotch the idea that determinism abolishes meaning by stating that thoughts and desires are really nothing more than neurons spitting signals through the brain. Consider the analogy of language.[14]

Are words real? When I write the word 'cat', you have a good idea of what I mean, because you have met real cats and learned to talk about them. When Tony Blair, then British prime minister, said of Islamism: 'What we are confronting here is an evil ideology', his listeners assigned some meaning to that phrase, in a way they would not have done if he had said: 'Blargy mella pilty shurp.'[15] Blair was referring to beliefs he did not share and did not like. I know this in the same way I know about cats, because the word 'beliefs' refers to things as real as cats: patterns of neural activity which bridge the gaps between stimuli and responses. These entities may not be objectively observable without advanced technologies, but neither is a stomach ulcer or an influenza virus. Nonetheless, all three phenomena are real.

Are words 'just' patterns of sound waves/written shapes/neural patterns/whatever? Of course not. These formats are necessary for us to be able to have and manipulate words, but the whole point about having correlations is that there are 'point-outable' things with which they correlate: objects, events, people's behaviours, you name it. To understand what a word means is to understand how people use that word. To understand what a neural pattern correlates with is to understand how that pattern functions; without reference to these causal interconnections neural activity might as well be random noise. But that activity is as real, as pointable-to (with the tools of modern neuroscience), as a word on the page. Both are meaningless unless we understand how they are used, what aspects of existence they go together with. Put in their context, however, constant and nearly constant conjunctions of neural behaviour with stimuli and responses—beliefs, desires, ideas, intentions, and so on—are both real and meaningful.

THE PROBLEM OF MORALITY

What, then, does neuroscience make of the notion that determinism undermines moral responsibility? The very term 'causal chain' is oppressive, is it not? In comes the stimulus, marching through your peripheral nerves like a dictator's army; conquest is a brief inevitable surge from synapse to synapse, and you twitch in response. Let us set aside the fact that, as a picture of what brains do, this is ridiculously oversimplified even if applied to your fastest reflexes, those which never get beyond the spinal cord. Consider an alternative: you, the authentic individual, are strolling contentedly along the street when—wham!—your arm moves without being caused to do so. Does this make you a truly free human being or a deranged flailing one? Surely freedom involves being able to act by choice. In other words, we want to act *for reasons*, not because the gods are capricious or the quantum leapt or the laws of physics suddenly took a vacation.[16]

Reasons, of course, are things that people—not brains—possess. They are also causes, not by-products, of mental life: they influence us to act in particular ways. Look at the language again: Rav goes to the shops *because* he wants to buy bread; Kris is jealous of Sam *because* Sam is dating Susie; Alice flinches at the sight of a thread on the carpet *because* she takes it for a spider. In other words, there has to be some causal link between a reason and the actions it influences. (If you had reasons to act, but they had no influence on your behaviour, would you be free? No, you'd be a random puppet.) Brains provide the material, if you like, of which these causal links are built, just as sound waves provide the physical stuff of speech and actions flow through the medium provided by our muscles. So what? They have to be mediated by something. Does the fact that the something is meat, or pressure waves in air, in itself make our thoughts, actions, or words any less free?

One cause of the unhappiness people feel about such arguments lies in the compelling metaphor of the 'causal chain'. If every step from input to output is determined by the previous step, where is the room for free will? But as we have seen, this picture—of humans as puppets controlled by strings of neurons—is deeply misleading. For one thing, they cannot be described as 'strings' even in the very simplest of our reflexes; the 'strings' feed back on themselves, tie themselves in knots, get tangled up with a myriad other connections. Neural 'strings' are also not at all like ordinary ones. Instead of one continuous thread, we have segments (neurons) joined by probabilistic devices (synapses), like a long road containing many bridges, each of which may or may not be missing at any given time depending on the history and current levels of traffic on that road and on all the other roads with which it connects. Yet again, we have no certainties here. Finally, each 'string' is as it is because of you: what you've done, your genes, your diet, your proclivities and beliefs, the people you encountered on your passage through the space-time continuum, your childhood experiences, and so, endlessly, on. The self may be multiple, the brain a noisy neural marketplace, but the conflicting ideas, the urges which tug you in opposite directions, are nonetheless yours.

You are unique. You are in charge.[17] Much of what you do goes unrecognized by your conscious self, but 'you' are a good deal more than just your conscious self—your body, for instance. You can act for reasons that seem good to you; you can reflect on your behaviour, decide to act differently next time, predict what's coming, change your mind, choose to try a new food, shop on impulse. What more do you want? Do you really feel less free because your brain causes some of your behaviour without 'you' noticing? (Yes, your brain micromanages the workings of your pancreas. Would you not be considerably *less* free if you had to supervise your innards all the time? Diabetes, which requires its patients to monitor just one aspect of their internal function, is not known for its liberating impact on human existence.) Would you really be freer if some of your actions occurred without being caused by those annoying neural precursors? No, because they are just the means by which *you* cause your actions to occur. Without them your actions would not be free because you would not have caused them to happen. You would be like a schizophrenic who believes that he does not control his movements—except that your belief would be correct. That would be possession or anarchy, not freedom.

Arguing that people are influenced by many situational factors (group pressures, perceptions of dominance, even background colours and sounds and odours) is not the same as saying that these environmental variables sufficiently cause behaviour as deliberate as cruelty. But of course, comes the response, a person's personality, acquired beliefs, genetic background, physiological state, past experience, and so on all make their causal contributions to the outcome. They feel very different from 'you', that little battered scrap of consciousness floating like scum on the surface of an ocean. Yet they are all parts of you, artificially distanced by the language we use to describe them. That scum of yours has the power to reach down into the ocean, understand its currents, and sometimes even alter their strength and direction. It too is a part of the ocean, and 'you' are the whole. And what does all that amount to, if not us acting as agents, causing changes in the world—partial, limited, easily biased agents, but agents nevertheless?[18]

MORALITY—REDUCE, REUSE, OR RECYCLE?

We evolved morality, Darwinian theories suggest, because it was useful. That should make us pause before we ditch it, because something tried and tested over hundreds of generations may not be so easy to replace. Legal rules and social conventions cannot cover every eventuality, and the modern world has weakened much of their authority. Even perpetrators of violent crimes like rape and murder are quite likely never to face any legal punishment. They may be ostracized by their communities, but in today's fluid world that is not the deterrent it was, since moving between communities—and sloughing off a bad reputation *en passage*—is so much easier.

We need strong sanctions for bad behaviour because, as social animals, we face two inevitable problems. The first is that we must interact with other human (and non-human) beings. If we harm them they are likely to harm us, if they can—and if they cannot their friends and relatives, or our disapproving kin, may punish us on their behalf. The second problem is that we spend much of our lives depending on the goodwill of other people. Entirely vulnerable as babies, we acquire defences which illness and old age soon sweep away. Power dependent on forcing people to do our will cannot last; we must make them *want* to help us. Moral rules, restrictive though they may seem to determined individualists, do just that, building trust, mutual respect, and the shared identities which make your loved ones' welfare important to you. Being a valued group member—being cared for—provides a form of social insurance which can be hugely beneficial in times of weakness. As studies have repeatedly shown, feebler social networks are a risk factor for poorer mental and physical health, less contentment, and even earlier death.[19]

If there is no God, why be moral? The second strand of Sade's challenge seems enticing, but it overlooks a crucial flaw in his assumptions. If power over others could be made absolute and eternal, there would be no reason for the powerful to concern themselves with the weak. In practice, however, power is always insecure. People age, alliances shift, regimes fall. The tormenting possibility remains, whatever is done to eliminate it, that one day the powerful sadist may find himself dependent on the people he used to torture.

Morality deters many instances of cruelty. Yet basic morality, as we have seen, is conditional; not everyone always enjoys its benison. Empathy can be suppressed, otherization enhanced, until cruelty comes to seem entirely reasonable. Perpetrators do not lose their moral inhibitions altogether, though they may squash them temporarily. The volunteers persuaded by Stanley Milgram to administer (as they thought) electric shocks to an innocent person did not turn the dial with easy indifference; they sweated, protested, and suffered. The SS men who murdered Jewish children may have hardened into indifference, or sadism, over time, but their contemporary writings are full of complaints about what they are being asked to do. One report from an *Einsatzgruppe* member reports that some colleagues 'could not cope with the demands made on them'. The report goes on:

> Many abandoned themselves to alcohol, many suffered nervous breakdowns and psychological illnesses; for example we had suicides and there were cases where some men cracked up and shot wildly around them and completely lost control. When this happened Himmler issued an order stating that any man who no longer felt able to take the psychological stresses should report to his superior officer. These men were to be released from their current duties and would be detailed for other work back home. As I recall, Himmler even had a convalescent home set up close to Berlin for such cases. This order was issued in writing; I read and filed it myself.[20]

The author describes Himmler's order as an evil trick: 'which officer or SS-Mann would have shown himself up in such a way?'. Nevertheless, some individuals did opt for—and obtain—treatment. Sick and injured German soldiers were also cared for. The Nazis did not abandon morality; they simply narrowed its range.

Real human beings, as opposed to positively or negatively idealized ones, are capable of being both kind and cruel. If we see cruelty as a quality of a person's character, then evidence of their kindness must dilute our moral judgement of their cruelty. But if we accept that cruelty is about behaviour, then the person is as responsible for their cruel behaviour as for their kindliness. Many constructed moral systems, such as the codes set out by Christianity, insist on exactly this distinction, separating prosocial virtues from antisocial sins and people from the sins which they commit. (In practice, 'hate the sin but love the sinner' has to be one of the most widely flouted injunctions in history.) Such moralities also counteract the group-centredness of basic morality by awarding more credit for good behaviour to strangers than to ingroup members.

Even psychopaths, for whom no moral code grips very deeply, are aware that moral codes exist for others. Certain behaviours are considered wrong and risk punishment. This knowledge does not disappear in situations conducive to atrocity. It may be deliberately ignored or reluctantly set aside, but the neural patterns which ground it have not stopped existing just because other desires have drowned them out. Morality is undoubtedly imperfect, but for evolved creatures imperfection is always where we begin the long attempt to change ourselves. Understanding the human basis of moral judgement allows us to begin to address its extensive problems. That, not abolition, is a realistic goal for scientific research.

As a modern, secular Westerner trained in science, I have assumed throughout this book that scientific understanding is both worth pursuing and applicable to any field of enquiry, even cruelty. My final task is to consider a potent challenge to that assumption. The challenge is not itself either scientific or philosophical; it is moral. It says that we cannot and should not attempt to explain human evil in scientific terms.

HERE THERE IS NO WHY

> ... what dreadful agony the sufferer must have laboured under, by being so frequently put to the torture. Most of his limbs were disjointed; so much was he bruised and exhausted, as to be unable, for weeks, to lift his hands to his mouth; and his body became greatly swelled from the inflammation caused by frequent dislocations. After his discharge he felt the effects of this cruelty for the remainder of his life, being frequently seized with thrilling and excruciating pains, to which

he had never been subject, till after he had the misfortune to
fall under the merciless and bloody lords of the inquisition.

(John Foxe, *Book of Martyrs*)

The claim is simple: in understanding cruelty we dissolve morality,
condoning atrocities by even attempting to explain them. By making
perpetrators seem less evil, we belittle the suffering they have caused
their victims—and this is not a morally neutral activity, however much
we may try to pass it off as impartial science. 'Here there is no why',
in the famous phrase from Claude Lanzmann's film *Shoah*.[21] The worst
atrocities defy our understanding. They are sacred, albeit at the negative
pole of sacredness, and should be kept out of science's grubby hands;
they are objects of awed and horrified incomprehension, abominations
which warn us away from wickedness. Their perpetrators have crossed
the line into evil. To say anything else is to desecrate the memory of
those who died.

SACRED TORTURE: ATROCITY AS EVIL

This takes us back to the essentializing of evil, to seeing human cruelty
less as the callous imposition of power (for reasons which might seem
reasonable to us) and more as malevolent sadism (the moral nadir of
sheer delight in torture and destruction). As we have seen, there are
problems with this approach; but before we review its difficulties let us
try, if we can, to understand why it gives comfort to its proponents. One
reason surely has to do with their personal suffering. Whether they were
victims themselves, lost loved ones, or saw some holy ideal attacked,
something of great value, a huge part of their symbolic (and in some
cases physical) identity, has been mutilated or torn away altogether. The
source of such pain will be immensely meaningful to these wounded
selves. To deny its meaning in effect denies the reality of their anguish.
And the cold amoral language of science, as many see it, has no place for
meaning.

Another benefit of sacralizing cruelties done to you, but not to others,
is that it can ease the weight of responsibility. Learning about so many
horrors can instil a weary, sickened guilt—even though you, now, can do
nothing to prevent them—making it preferable to focus on the pain you

and your group know best. Treating your agony as special can also make you disturbingly blind to your own less-than-kind behaviour to other people. Victims of atrocities can sometimes behave as if their trauma has given them a total moral discount for future cruelty, even against targets unconnected with the perpetrators who attacked them. This logic is uncomfortably close to that employed by terrorists: my grievances are so huge that they justify any measures I may take, even killing the innocent. (You can't punish me, my trauma made me do it.) Tempting though it may be to shift the heavy moral burden of being human—that is, the descendant of callous and probably murderous predecessors— onto a few clear targets, such myopia can make damaged people cruel.

Many victims, however, are so sickened by the cruelty they experienced that they reject all further violence, even against their torturers. For them perhaps what matters is that evil is by its nature meaningful. Evil-doers, unlike senseless chance, have agency and choice; they desire the victims' suffering, which makes it important to them. Victims can see their appalling agony as having an appropriately potent and significant cause: the fully intended malevolence of the perpetrators.

Alas, most human cruelty is callous or careless, not fiendishly sadistic. Setting aside the often unresolvable problem of how to determine perpetrators' intentions, sacralizing atrocities plunges victims into the essence trap. This is especially noticeable with genocide. The moral millstone attached to accusations of sadism is also what makes accusations of genocide so much more controversial than labelling the same events as massacres, since genocide, like sadism, fixes blame on a perpetrator's nature.[22] The desire to annihilate and the desire to make people suffer can be otherized as pure evil. While demanding action, this absolves us of the responsibility to take such cruelty seriously as changeable behaviour.

If our morality were truly victim-centred, in the sense of treating victims democratically as human creatures with lives and loves and dignity to lose, we would spend less time worrying about perpetrator motives and more time focused on repairing and preventing harm. Sufferers and the bereaved would not have to struggle, sometimes for years, to get their torture recognized. We would see the murder of any individual as equally abominable, and their human value as equally sacred, whether they were killed in a genocide, a terrorist attack, a pre-planned murder, or a drunken street brawl. Cruelty is certainly judged by the psychology of those who act cruelly, but it also includes the suffering of victims. However, our morality is not like this. Our care is uneven, victims are not all equal, and otherization can taint a victim as well as a perpetrator. Besides, our main concern when confronting extreme cruelty is not with what it has already done to other people but with what it may do to us. This, together with the glamour of evil, drives our obsession with perpetrators' mental states.

Nonetheless, however we may feel about victims of cruelty, we cannot let them insist that their trauma is unique. On the private and personal scale, of course, it is, because each victim is unique; but at the level of public understanding we have no justification for ranking one genocide, massacre, or spree killing as morally worse than another. And yet we do.

If you are a citizen of the West, your concept of genocide will have been heavily influenced by the Holocaust. You may thus think of genocide as something systematic, technological, bureaucratic; the deliberate attempt by a deranged leader and his coterie to exterminate a particular group of people. But Hitler's Final Solution was driven by the actions of individual Nazis on the ground as well as by high-level decision-making. Like all bureaucracies, it was riddled with local inefficiencies, petty competition, and special cases. As for technology, the Nazis are known for developing gas chambers, but they and their allies killed many of their millions of victims with bullets, labour, torture, hunger, and disease. The regime of Democratic Kampuchea, lacking Zyklon B, murdered up to 2 million people in four years (an average of around 1,000 a day, in the ghastly amoral calculus of such estimates), frequently by beating them to death. Perpetrators in the Rwandan genocide took a mere hundred days to kill at least 800,000 people, and they used mainly guns and machetes.

Nor was the Holocaust the first genocide, though the term was coined in response to it.[23] Julius Caesar wiped out entire tribes, killing many and selling the rest into slavery. The Holocaust was not even the first modern German genocide. German colonialism opened the twentieth century by ravaging—one cannot say 'decimating', the losses were closer to eight in ten than one in ten—the Herero people of what is now Namibia.[24] Germany was not the only European perpetrator. Belgium, for example, wreaked havoc on the inhabitants of the Congo.[25]

So is the Holocaust, as some people argue, uniquely evil?[26] Really? Do we honestly want to downgrade the horrors of what happened in Rwanda, Cambodia, colonial Africa, and the Americas? Do we even want to set 'genocides' apart from other human-caused or human-enhanced mass deaths, thereby downplaying what happened in China during the Second World War, the Ukrainian Holomodor, the Irish Famine, Indonesia in the 1960s, and many others?[27] These often-forgotten victims were just as likely as European Jews to suffer and to be mourned by their bereaved. And for raw sadism, 'lesser' atrocities can match the Nazis' depredations. Here is the American adventurer Walter Hardenburg, publishing details of what he saw in the Amazon's Putumayo region in the first years of the twentieth century, where a Peruvian corporation backed by British investors was using native (mainly Huitoto) Indians as forced labour for rubber extraction. Hardenburg observed:

> that the peaceful Indians were put to work at rubber-gathering without payment, without food, in nakedness; that their women were stolen, ravished, and murdered; that the Indians were flogged until their bones were laid bare when they failed to bring in a sufficient quota of rubber or attempted to escape, were left to die with their wounds festering with maggots, and their bodies were used as food for the agents' dogs; that flogging of men, women, and children was the least of the tortures employed; that the Indians were mutilated in the stocks, cut to pieces with machetes, crucified head downwards, their limbs lopped off, target-shooting for diversion was practised upon them, and that they were soused in petroleum and burned alive, both men and women.[28]

The exaggerations of a maverick character? Not this time. The British government, which had shares in the corporation responsible for the rubber-gathering, sent its consul Roger Casement (who also reported on atrocities in the Congo) to investigate the Putumayo claims. His report, which caused an outcry when it was published by the British government in 1913, estimated that twelve years of white oppression had reduced the population by around 30,000.[29] Yet who now remembers the Putumayo? The killings were not genocidal, just another instance of colonial oppression. The corporation seems to have thought the supply of slave labour would be endless, leading it to hold individual lives extremely cheap. Out of this callous cruelty emerged the sadism which so shocked Hardenburg, Casement, and their readers.

Is it worse to kill someone because they symbolize a human type whose existence you have decided to find intolerable than because they stand between you and easy gain, or are a drain on your resources, or simply because there are lots more where that one came from? In the first, genocidal case victims are the perpetrator's obsession; in the others they are at best a nuisance. No wonder victimhood seeks the label 'genocide'. Like 'evil', it bestows significance, helping people come to terms with the worst human existence has to offer.

HERE THERE MUST BE WHY

How are researchers who try to understand atrocities to react, faced with this passionate distress? Some say that individual pain is part of the moral and subjective world, whereas science deals with objective reality, and must be value-neutral if it is to succeed. This response invokes the naturalistic fallacy, that famous scientific defence mechanism which distinguishes 'is' (facts) from 'ought' (values) and insists that the two be kept apart. Scientific explanations are not moral justifications, the argument goes, so scientific theories of cruelty are irrelevant to moral responses.

But are they? Scientific approaches can do much to clarify and interpret patterns of cruel behaviour, but cruelty also has moral aspects, including agency and responsibility. Crucially, these are affected by our perceptions of causal factors as well as moral ones (the murderer with a damaged prefrontal cortex gets a moral discount compared to his healthier cell-mate). The border between science and morality is not hermetically sealed, in other words, because scientific data can alter moral judgements. Morality, whether basic or constructed, is not the primitive, impervious reflex that its ancient origins might lead us to believe; accumulated layers of control have added flexibility and nuance.

Explanations of atrocities may not themselves *be* moral justifications, but they can *influence* them; moral judgements change as knowledge changes. Scientific theories of cruelty may thus have serious moral consequences, however much some researchers might wish they didn't.

There are other counter-moves to the objection. One is that empathy has inescapable limits. I can never feel your agony as severely as you do, so it will never matter quite as much to me, or anyone else, as it does to you. Humans are just built that way. What I can do is understand that you are suffering and that your suffering is real and meaningful. That is, I can respect it and agree that it has moral weight.

Another response is that, just because you feel your anguish more than I can, that does not mean I feel no pain at all and do not care. Researchers who study human cruelty may have various motivations, including empathy for victims. Some have lost friends and family in atrocities. Yet the belief that understanding human cruelty will help us to reduce it should not require additional defence. Victims of atrocities and their friends and families have a voice, an authenticity, which no science can match; but expressing agony cannot be the purpose of scientific study—and emotions by themselves do not help us understand why atrocities occur.

A cry of pain can highlight some vile misdeed and give it meaning; and that, in our preoccupied world, is an essential function. It is easy to look away and forget. Recall the Rwandan genocide, with its staggering death-toll: averaging more than one Srebrenica, nearly three 9/11s, every day for over three months.[30] Yet I recently witnessed two of the most accomplished senior scientists I know—highly educated, liberal-minded people—unable to remember even the names of the factions involved. 'Tutus, or whatever', said one, moving hastily on. Such ignorance, not uncommon, highlights the continued need for cries of pain.

What remembering and grieving and swearing 'never again' cannot do, however, is translate benevolent impulses into the effective prevention of human cruelty. One of the most tragic lessons of history is that moral impulses, so carefully inculcated in culture after culture, seem almost irrelevant when it comes to committing atrocities. Christianity teaches love for others; Islam tolerance and peace; Communism equality and social justice; and the modern West democracy and freedom.

All of these belief systems—like others not listed—have followers who kill atrociously, horrors committed in the name of high ideals. Clearly, moral prohibitions don't stop cruelty. Understanding why it happens won't stop it either, but it is our most likely route to reducing cruel behaviour.

Summary and conclusions

The moral codes which reward and punish us evolved to help us act in ways which bring us long-term benefit and not act in ways which do us long-term harm. They do so by laying down the neural patterns which serve to inhibit our cruelty and boost our kindness—sometimes. Now at last we are starting to understand the deeper reasons which make our moral judgements at once so compelling and so limited. That awareness gives us the power to change them. For instance, we can use the notion of moral discounting to see responsibility not as all-or-nothing, but partial and mediated, judging that some factors (like certain forms of brain damage) offer large discounts, while others (like obedience) do not. We can decide to extend our moral universe to cover our species, or even other species, as evidence emerges of their many similarities to us. We can even attempt to free our moral beliefs of the dangerous pull towards essentializing and sacralizing good and evil, seeing cruelty more as a matter of human weakness. In doing so, we may be better able to accept responsibility for the imperfections which lead us to moral failures.

What we cannot do is abandon morality altogether, any more than we can decide not to be the kinds of creatures which breathe and defecate, lust and grieve, desire power and crave affection. Sadeian nihilism is not a practical option for as long as we are susceptible to reward and punishment and vulnerable to the power of others. As for the claim of moral relativism, that no morality can pronounce itself superior to others, that move is suspect. Some moral codes clearly fit us better than others (utilitarianism, outright selfishness, and total altruism, for example, seem not to fit us particularly well). Our morals are a mixed bunch, adapting to suit the circumstances: callous otherization when we feel at risk, loving-kindness when we have learned to care.

We could attempt to reduce the triggers for cruelty, for instance by querying those who warn us of symbolic threats and challenging their dubious ideologies. Education and the redistribution of political power offer hope for improvement. We could heighten the costs of war, in which cruelty flourishes, for those who instigate it. But the underlying mechanisms remain, part of our neural inheritance. Cruelty is often the easiest option, and being cruel can be exhilarating. We must accept that and learn to live with it—if we can.

Failures of empathy, selfish disregard for others, easy hatred, and hideous cruelty are just as human as language, love, and heroic self-sacrifice, which can also make us cruel. Those of us who have not yet become victims or perpetrators of violence can thank historical luck, not our own innate marvellousness; being decent offers little protection. Meanwhile, the need for control has created apocalyptic weapons which could turn one group's cruelty into global catastrophe. Cruelty may consume us before we can master it.

How much time is left to us? Will *Homo sapiens*, that extraordinary species, grow up to live with less torture and abuse, or will our governments use the science of cruelty to make their fighters and leaders more callous, their citizens more apathetic? Will perpetrators continue to gamble on bystander indifference and win, or will observers give their protests political force? Arguing that cruelty is a moral concept does not mean we should not take it seriously. Far from it. We are moral creatures and could take it more seriously than we do, treating cruel behaviour not as devilishly beyond us but as something we can all do something about, be it only protesting to friends, the media, and governments about specific instances of cruelty. We are symbol-users, able to change our behaviour and other people's once we understand it, using the tools of politics and science.

We cannot remove human cruelty entirely. It is too useful and, at times, enticing. We could, however, reduce it if we chose, making the cultural changes which would render callousness less socially acceptable. But that would take effort, and so far that effort has not been forthcoming. Instead, political leaders have fostered cruelty and used it to their advantage, with the tacit or explicit consent of their supporters. Now that science can offer us the chance to understand and limit cruelty, the risk is that such knowledge will be abused in order to make some of us—soldiers, for example—still more cruel. Yet humankind still has time to make different choices. We could make people less cruel, if we cared enough to do so.

Will we do it? My own opinion is that any changes we manage to make will be minor, slow, and grossly inadequate to the task at hand. Cruelty is terrible, but for most of us it is not at present terrible enough. Those who have the power to make a sizeable difference have built themselves mighty defences against attack from others. Cruelty does not affect them directly, and so it is not as real to them as it is to those who suffer or perpetrate it. The powerful, having done enough to keep the vicious corralled into distant ghettoes, feel no pressure to take any further action. Even should they lose political power, they cannot imagine being forced to live as victims live. President Richard Nixon was not rehoused in the Vietnam jungle carpet-bombed by forces under his command, nor even in one of the more lawless neighbourhoods of Washington, DC. Insulated from the effects of cruelty, is it likely that he—or any of us—should care? Yet no political power can last for ever. One day we may be forced to care.

People are not entirely selfish. Moral passions can lead to social change, and in many respects our world is much less cruel than it was. Nonetheless, even highly educated, leisured, well-fed, and well-protected human beings continue to display atrocious cruelty at times. Understanding why people are cruel requires intelligence, of which human beings have a plentiful supply. It is a challenging task, but not impossible; if intelligence and knowledge were all that were required we could begin ensuring less cruelty tomorrow. Yet knowledge is not enough, and no amount of clever thinking by delegated experts can help us make the necessary changes. For that we need wisdom, courage, and the will to exert considerable effort without much discernible reward. These qualities are hard to find in the human species. Lacking them, we remain at risk: of suffering cruelty and of being cruel.

Notes

Introduction: Cruelty in context

1 See Boklage (1990). A large-scale study from the US Centers for Disease Control and Prevention reports that, in 1996, 'An estimated 6,240,000 pregnancies resulted in a live birth, induced abortion, or fetal loss in the United States'. Of these, 980000 (16%) resulted in fetal loss; 62% survived to term and 22% were aborted (Ventura et al. 1999).

2 See Nuland (1994) for more on the basic mechanics of common kinds of death. With respect to violently inflicted deaths, any statistics cited should be treated with caution, especially when they refer to large-scale massacres. A major source for these numbers, R. J. Rummel's *Death by Government* (1994), describes them as 'fundamentally nothing short of wrong' (p. xviii), which is unsurprising given that record-keeping is not normally high on perpetrators' to-do lists. The numbers are given to provide some sense of the magnitude of the destruction involved.

3 Over a lifetime, a person's estimated likelihood of being murdered varies widely across cultures, from 1 in 2,000 (United Kingdom) and 1 in 200 (United States) to greater than 1 in 20 (Colombia, South Africa) and even higher for some tribal cultures, such as the notoriously violent Yanomamo (Buss and Duntley 2005).

4 The US Department of Justice, summarizing national homicide statistics (US Department of Justice 2007), states that, in 2005, 'Males were almost 10 times more likely than females to commit murder'. The gender more likely to perpetrate violence is also the gender more likely to suffer it: men are almost four times more likely to be murdered than women and also more likely to die by accident and suicide (Day 1984).

5 Fisk (2006).

6 The book featuring the brain-eating psychopath is *Hannibal* (Harris 1999). The film, released in 2001, was directed by Ridley Scott.

7 Some thinkers argue that the twentieth century's atrocities, such as Auschwitz and Hiroshima, were of such magnitude that they demand a full-scale rethinking of Christian theology. See e.g. Fasching (1992); also Shklar (1984), 7–44.

266

8 Sade (trans. 1991), 698.

9 A useful discussion of cruelty and evil can be found in Berkowitz (1999).

10 *Guatemala: Memory of* Silence, Report of the Guatemalan Commission for Historical Clarification (Comision para el Esclaracimiento Historico) (2007).

11 Peter Kürten, surely a defining exemplar of the modern sexually sadistic serial killer (Berg, trans. 1938), was tried and executed in Germany in 1931. The number of his victims is unknown. Ian Brady, assisted by his girlfriend Myra Hindley, is known to have killed five children in northern England in the mid-1960s (Williams 1967).

12 Rees (2005), 8.

13 Throughout this book I will focus on active cruelty, although cruelty due to neglect (when people with responsibilities for care choose not to fulfil those responsibilities) can also be considered within the framework developed here.

14 See e.g. Beck (1999), Staub (2003), Sternberg (2005), Waller (2002), also Gregory Stanton's model of genocide (Stanton 1998) and the related concepts of tribalism (e.g. Glover 2001) and infrahumanization (Castano and Giner-Sorolla 2006, Cortes *et al.* 2005; see also Ch. 2, n. 23).

15 Otherization appears to begin early in life (Kinzler *et al.* 2007).

16 Actions and emotions are not merely products of brains designed to think. Rather the reverse: rational thought looks more like a by-product of the brain processing than its central *raison d'être*. See Maxwell and Davidson (2007).

17 Mackay (ed. 1973), 10.

18 See e.g. Sharot *et al.* (2007), Taylor and Gollwitzer (1995); also Malle (2006).

19 Valdesolo and DeSteno (2007).

20 A standard opposition in psychology is between essentialism and situational-ism. The idea that people have some core essence which determines their personality has been heavily challenged by social psychologists, who argue that far more of our behaviour than we like to think is influenced by what happens to be going on around us (much of which we may not notice). In practice, many researchers adopt a position somewhere between the two poles, in which individuals (or groups) can have differing predispositions to act in certain ways, for example due to genetic variation or differing moral values, without these necessarily being immutable.

21 Distance in time and geographical distance can make people more prone to slip into the essence trap (see Nussbaum *et al.* 2003; also Bandura 2002).

22 Browning (1991).

23 There are many books on the Rwandan genocide. I have referred primarily to Gourevitch (2000), Dallaire (2004), and Hatzfeld (trans. 2005). My main source for information about the turmoil following the break-up of Yugoslavia has been Noel Malcolm (1996; 1998). See also Ch. 9, n. 30 on Srebrenica.

24 Aeschylus (trans. 1999), Euripides (trans. 1963).

25 Klee *et al.* (1991), 42.

26 Kelemen (2003).

27 Some atrocity stories are even true, like the rumours which rippled out from the Nazi death camps of children torn from their mothers' arms, corpses robbed of gold teeth, human bodies used to make soap. Ironically, they were widely disbelieved at the time (in some quarters they still are). The point is not that atrocities do not happen, but that stories about them can serve other functions as well as simply communicating information (Mertus 1999).

28 Frankfurter (2006), esp. 6–12.

29 Ibid. 11.

30 Fein (1990).

31 Semelin (trans. 2007).

32 'Interdisciplinary', one of the most lip-serviced terms of academia, need not mean 'indiscriminate'. This book, for example, excludes or merely mentions a number of areas, from psychoanalysis to game theory to literary criticism, which might be considered relevant to its topic. For an example of how focused interdisciplinary research can usefully address the issue of atrocities in war, see Kassimeris (2006).

33 Functional magnetic resonance imaging (fMRI) measures changes in blood-flow in a living brain as the volunteer performs a task, or simply rests. The changes are very small (a few percent), and the statistical manipulations required to analyse fMRI data are extensive.

34 If you are feeling short-changed by all this uncertainty, incidentally, you have probably been misled by the numerous commentators who talk of 'science' as if it were all some nearly completed monolith with the intellectual bulk and predictive prowess of, say, quantum mechanics. This is not the case. Neuroscience is nowhere near as solid as modern physics, for two reasons. First, the objects being studied are more complicated than anything in physics, even the tangled webs of string theory or the mathematics of complexity. Secondly, physicists have had a lot longer to make progress. They date back at least as far as the ancient Greeks, when expert opinion saw brains as radiators (Aristotle, trans. 1937, *Parts of Animals* II. vii). I caricature the mighty Aristotle, but the point remains that neuroscience only really began to develop as a science long after his observations of the brain. Youth brings exuberance, and neuroscience is excitingly exuberant, but settled stature and power are some way off yet.

35 Tilly (2003).

36 Research emanating from affective, behavioural, and clinical neuroscience paradigms is converging on the conclusion that there is 'a significant neurological basis of aggressive and/or violent behavior over and above contributions from the psychosocial environment' (Bufkin and Luttrell 2005, 187).

37 The syndrome of psychopathy is described in Hare (1999).

38 The field of research on aggression, conduct disorder, and psychopathy is gigantic. Here are some pointers: Anderson *et al.* (1995), Blair *et al.* (1997), Dadds *et al.* (2006), Frick and Dickens (2006), Kramer *et al.* (2007), Krueger *et al.* (2007), Lindsay and Anderson (2000).

39 The psychological study of aggression has led to the development of the widely accepted General Affective Aggression Model (Anderson *et al.* 1995, Lindsay and Anderson 2000), which has recently received support from neuroimaging (Kramer *et al.* 2007). I will not be discussing this model in detail, primarily for reasons of space, but also because this is a book about cruelty, not aggression. I have chosen to focus instead on a less well-known contributor to hostile behaviour: disgust (see Ch. 5), which is particularly relevant to cruelty. However, the brain networks involved in mediating reactions to disgust overlap extensively with those mediating aggressive responses, and much of what I say about disgust is relevant to the GAAM (particularly the discussion of anger-threat responses in Ch. 3).

Chapter 1: What is cruelty?

1 Klee *et al.* (1991), 31–32.

2 The *Oxford English Dictionary*, for comparison, traces 'aggression' and 'attack' to the seventeenth century, 'barbarous' (in the sense of cruel) and 'atrocity' to the sixteenth, and 'violence' to the thirteenth century. 'Sadism' was first recorded in English in 1818, 68 years before the great clinician of sexual deviance Richard von Krafft-Ebing published his *Psychopathia Sexualis* in 1886 and only four years after Sade's death in 1814 (Krafft-Ebing, trans. 1965).

3 Champlin (2003), DeBoer and Maddow (2002). Champlin discusses Tacitus' description of Christians being used as torches (pp. 121–2).

4 *The Iliad*, VI. 63–70, translated by Robert Fagles (Homer, trans. 1999). For more on the poem's dating see Bernard Knox's Introduction, p. 5 ff.

5 A useful comparison is Claudia Card's analysis of the concept of evil (Card 2002), to which I owe much.

6 There is evidence that during the war the Germans encouraged or even manufactured some apparently spontaneous displays of anti-Semitism. A report from the leader of *Einsatzgruppe* A (see Ch. 2, n. 35) says that 'local anti-Semitic elements were induced to engage in pogroms against the Jews ... The impression had to be created that the local population itself had taken the first steps of its own accord as a natural reaction to decades of oppression by the Jews and the more recent terror exerted by the Communists' (Klee *et al.* 1991, 24). The commander notes that getting large-scale pogroms under way 'was initially surprisingly difficult' (p. 27), but it was achieved, and not by explicit coercion. Instead, Lithuanian partisans were recruited with the incentive that cooperation might gain them power in a future government. Needless

to say, there is a gap between being offered incentives to behave in a certain way and being forced to do so.

7 The concept of cruelty as excessive punishment can be traced back at least to the Roman writer Seneca's essay *De Clementia*, which discusses mercy as shown (or not) by those in power (Seneca, trans. 2007). The history and legal implications of this view are discussed in Barrozo (2008).

8 The 'laws of war', whose ancestry lies primarily in the Christian 'just war' tradition, were formally set out in the Hague Treaties of 1899 and 1907 and the Geneva Conventions of 1864, 1929, and 1949. See Sorabji and Rodin (2006) for an interdisciplinary discussion of the ethics of warfare, Howard *et al.* (1994) for a more historical approach.

9 Note that the issues of justification and desert are not identical. Justification has to do with the perpetrator, desert with the victim. Imagine, for example, that a man has been accused of sodomy, apparently by eyewitnesses, in a country which punishes such behaviour by death. According to the prevalent moral conventions, he deserves to die. Yet those who accuse him would not be justified in promptly killing him. Not only must the case be tried (the eyewitnesses might be lying), but the sentence is the prerogative of the state, not of individual vigilantes. One could also imagine more abstruse scenarios. For instance, a perpetrator could falsely accuse his victim of one crime in order to justify inflicting suffering—without knowing that the victim has indeed committed another crime for which, should he be tried, he would receive equivalent punishment. In that case the perpetrator could not honestly claim that his behaviour was justified, but the victim would nonetheless deserve to suffer.

10 See also the discussion of the *Oxford English Dictionary*'s definition of cruelty in Kekes (1996). Kekes amends the definition by asserting that the harm should be inflicted 'in a way that endangers the victim's functioning as a full-fledged agent' (p. 837), but says nothing about how this is to be assessed or why it is his preferred criterion. It has its problems; how, for instance, is cruelty to babies and the mentally disabled (not usually seen as full-fledged agents) to be viewed on this definition? Since my analysis relates to everyday rather than to philosophical concepts of cruelty, since everyday language allows the term 'cruel' to be used for more 'minor' kinds of cruelty, and since perceptions of 'endangered functioning' may vary between the victim, perpetrator, and third parties, I will set the issue of defining 'serious' cruelty aside, as raising more problems than it resolves.

11 For a discussion, and neurological example, of the difference between 'pain' and 'suffering', see Damasio (1996), 262–6.

12 A specific gene whose mutation impairs pain perception has been identified (Cox *et al.* 2006, Goldberg *et al.* 2007). Such individuals appear to be able to feel empathy, to some extent (Danziger *et al.* 2006).

13 See e.g. Alicke (2000), Borg *et al.* (2006).

14 A classic experiment on perception of agency was done by Fritz Heider and Marianne Simmel in 1944 (Heider and Simmel 1944). They used a video of three small moving geometric shapes (a circle and two triangles), and found that participants readily interpreted the movements as purposeful actions. For an updated view of the research see Heberlein and Adolphs (2004). See also Dennett (1989), Gauthier *et al.* (2000), Gergely *et al.* (1995), Ulloa and Pineda (2007); and for recent research on anthropomorphism see Epley *et al.* (2007).

15 Krumhuber *et al.* (2007), Simion *et al.* (2008). The sense of agency can also affect our perceptions, even when consciousness is not involved (Maruya *et al.* 2007). Research further suggests that the human bias towards detecting living things is not due to expertise (New *et al.* 2007); i.e. it may be an evolved response rather than a learned one.

16 For more on the causal thinking to be found in other species, see Dennett (1989), Hauser (2006).

17 An example of a perpetrator's inability to explain his behaviour, even when he is openly remorseful, is given by the Japanese war veteran who admitted to having killed more than 200 people: 'There are really no words to explain what I was doing. I was truly a devil' (Chang 1998, 59). His murder-toll is shocking, but it is a drop in the ocean of carnage resulting from Japan's wars of conquest in Asia. Civilians and prisoners of war in China and other occupied countries suffered horrific abuse (see e.g. McArthur 2006). Perhaps the best-known atrocities were committed during the Rape of Nanking (now known in the west as Nanjing), during which more than 260,000 people (some sources say more than 350,000) are thought to have been slaughtered over seven terrible weeks. The conservative figure (acknowledged as such) is from the official war crimes trial authority, the International Military Tribunal for the Far East.

18 Nell (2006). A common distinction is between motives—consciously recognized reasons for actions—and the causes which facilitate a particular action, which may or may not be recognized by the actor. The mismatch between the factors which influence us and our awareness of them contributes to the complexity of moral judgements.

19 The philosophical process of defining terms is immensely useful, in that it clarifies underlying concepts, but it can display two somewhat problematic tendencies. The first (generic to academics) is to privilege the technical definition over everyday usage even when the two are quite distinct, in effect telling 'ordinary people' that their term has been colonized by specialists, altered, and given back to them with instructions to sharpen up their usage accordingly. The second is to present the definition as if it has a neat and discoverable boundary, such that for every case of potential cruelty the truth of whether it is or isn't cruel is, in principle at least, discernible and absolute (i.e. independent of who is doing the

discerning). Both tendencies, it seems to me, assume a rather digital view of language and brain function (though this may be because I was made to read Wittgenstein at an impressionable age). In practice although some words can be precisely defined, many have meanings which blur and shift around the edges (what does 'red' mean to you?) and some, like cruelty, have highly contested meanings. Fuzzy blobs rather than tidy packets is certainly what our understanding of neuroscience, with its emphasis on probability, suggests we should expect. I am therefore uncertain as to whether, for cruelty, the high road of abstraction would lead us anywhere useful in the end. *Cruelty*'s concern is in understanding how cruelty is generally conceived—which may or may not be philosophically inconsistent—rather than in attempting to untangle any logical knots and achieve a precise definition of the term.

20 Hinton (2004), 47.

21 Majdandzic *et al.* (2007).

22 Merz-Perez and Heide (2003), 109.

23 See e.g. Darby and Jeffers (1988), Devine *et al.* (2001), Dumas and Testé (2006), McKelvie and Coley (1993).

24 Greene *et al.* (2004). The taxonomy of two moral systems echoes distinctions made by many philosophers. See for example Thomas Nagel's discussion of absolutist and utilitarian approaches to moral challenges in warfare (Nagel 1972).

25 Bentham (ed. 2007).

26 Thagard *et al.* (2006).

27 Greene *et al.* (2004).

28 Gailliot *et al.* (2007).

29 Thagard *et al.* (2006).

30 Vaish *et al.* (2008).

31 Sade and Nietzsche, for example, both railed against the dominant morality of their time as, in Nietzsche's words, 'a piece of tyranny' (Nietzsche, trans. 1973, 92) in which 'everything that raises the individual above the herd and makes his neighbour quail is henceforth called *evil*' (p. 105).

32 Hauser (2006).

33 de Waal (1996*a, b*), Hauser (2006).

34 Chomsky (1957), Hauser (2006), Mikhail (2007). For a critique of the 'moral grammar' approach see Dupoux and Jacob (2007).

35 It is not unknown for people to argue that, for example, male jealousy is natural ('has an evolutionary basis') while shame is 'cultural' (i.e. less securely grounded in biology); *and therefore* jealousy is a more authentic/genuine/legitimate emotion than

shame. This is poor logic. It is also dangerous, if used to imply that the often lethal behaviour of jealous males can be morally justified.

36 Kindness and hospitality to strangers, of the kind promoted in, for instance, ancient Greek traditions (e.g. Homer), is not incompatible with wariness. The relevant principle is 'tit-for-tat'. Kindness to an apparently unthreatening stranger may open up opportunities for mutually beneficial intergroup interactions, such as trade. For as long as the stranger does not show hostility, being nice is an efficient strategy to pursue. If the stranger reacts with hostility, however, that is seen as social betrayal and punished severely. Similarly, killing the stranger who had accepted one's hospitality was regarded as 'a crime of mythical proportions in Greek tradition' (Frankfurter 2006, 77).

37 The philosopher Peter Singer has shown, for instance, that people may readily agree that some moral problems, such as the presence of terrible poverty alongside great wealth, should be addressed, as a matter of moral obligation. The same people refuse to give up any of their own excess wealth to help relieve poverty (Singer 1971).

38 Keeley (1996).

39 Coates (1997) discusses just-war theory, including the proportionality criterion.

40 For a recent discussion of Athenian democracy, see Raaflaub *et al.* (2007), esp. 11–12.

41 Many commentators, from left-wing veterans like Noam Chomsky to international organizations like Amnesty, have criticized the US government's handling of prisoners captured as part of the 'war on terror', particularly at the Abu Ghraib prison in Iraq and the Guantanamo Bay camp in Cuba.

Chapter 2: Quis judicat? Who decides?

1 See e.g. Sampson (1993).

2 The Stanford Prison Experiment, as Philip Zimbardo notes in his book on the subject (Zimbardo 2007, 235), was instrumental in changing the ethical climate for psychological research. Zimbardo reports that he had the experimental ethics specifically evaluated by the American Psychological Association in 1973. Since then much has changed; nowadays the experiment would not be ethically acceptable.

3 Baumeister (2001). This is a specific instance of the 'Rashomon' effect, named after the film by Akira Kurosawa in which the same story is told from the perspectives of more than one participant. The film is based on a short story, 'In a Bamboo Grove', by Ryunosuke Akutagawa (Akutagawa, trans. 2007).

4 See e.g. Boden-Albala *et al.* (2005), Cacioppo and Hawkley (2003).

5 Recent books on the topic of why consumer capitalism has made us wealthier but no happier include de Graaf *et al.* (2001), James (2007), and Offer (2006).

6 Louis P. Lochner's 1943 book *What About Germany?* (Lochner 1943) brought Hitler's speech of August 1939 to public attention, including the notorious rhetorical question: 'Wer redet heute noch von der Vernichtung der Armenier?' Whether the mass killings of Armenians in 1915 should be considered genocide is a hugely controversial issue, especially in Turkey. (Incidentally, the Turks, like Hitler, appear to have been inspired by Genghis Khan: Dadrian 1995, 403–9.) The key word in Hitler's question is *Vernichtung*, which carries the same sense of changing something to nothing (*nicht*) as its English counterpart annihilation (from the Latin *ad nihil*, 'to nothing'). The Nazis used *Vernichtung* to describe the Holocaust, and four decades earlier Germans used the same term for their slaughter of the Hereros in Africa, which reduced the Herero population from around 80,000 to 20,000 (Bridgman and Worley 2004, Gewald 1999). Is *Vernichtung* then equivalent to genocide? No, because genocide is a post-war legal construct requiring the provable intention to destroy members of a group purely because of their group membership. This intent is difficult to prove in a court of law, as perpetrators are often careful to avoid providing written orders (for instance) which could be used against them. Even in cases widely accepted as genocide, like Rwanda and the Holocaust, very few of those responsible have been brought to justice. (Some Turkish leaders were tried and executed after the First World War.)

7 Research using twins suggests that the sense of fairness is subject to considerable genetic and only modest environmental influence (Wallace *et al.* 2007).

8 Nagel (1972).

9 See e.g. Browning (1991), Milgram (1997), or Newman and Erber (2002).

10 Goldhagen (1997).

11 The very fact that the phrase 'run-of-the-mill atrocities' makes sense in the way that, say, the phrase 'dark sunlight' does not surely tells us much about the human capacity for moral disengagement.

12 Baumeister (2001).

13 The furore over the 'Twinkie defence' arose because commentators thought that the defendant's lawyer had made a claim which was not in fact made: that the defendant's consumption of junk food had affected his brain and therefore diminished his responsibility for his behaviour. The actual claim was that the defendant was suffering from depression which diminished his responsibility, and that his consumption of junk food was evidence of his depression (he had previously eaten healthily). See Mikkelson and Mikkelson (2007).

14 Shakespeare (ed. 1997).

15 Melson (1992).

16 Jacques Semelin's masterly *Purify and Destroy* (Semelin, trans. 2007) has much to say on this double-sided (both internal and external) aspect of a society's 'evil Others'.

17 Rees (2005), 139.

18 The Turkish quotation is from Morgenthau (ed. 2000), 223. Henry Morgenthau, US ambassador to Turkey in 1915, when the massacres began, did much to bring Armenian sufferings to public attention. Over a million Armenians died (Dadrian 1995). (Morgenthau's book, like many others on the topic, has of course been denounced as propaganda by Turkish commentators.) The second quotation is from Danner (2005), 75, and refers to the massacre at the El Salvadoran village of El Mozote, an event said to have been the worst single atrocity in the region's modern history. El Mozote was part of a so-called clean-up (*La Limpieza*) designed to starve rebel guerrillas of resources by destroying the villages which allegedly supported them. Irrespective of age, sex, or state of health, villagers were murdered by an elite battalion of El Salvador's army, commanded by Domingo Monterrosa. Over 500 dead were named; many more were not identified.

19 Abelson *et al.* (1998).

20 Tuol Sleng was the most notorious of the prisons in the Khmer Rouge regime of 'Democratic Kampuchea', which killed up to 2 million Cambodians (of around 7 million) during the 1970s. Over 14,000 were tortured and killed in Tuol Sleng; a handful survived. For an excellent examination of the DK nightmare, see Hinton (2004). See also Rummel (1994), table 1.2.

21 Examples of human biases include distorted perception of causation (White 2006), the assessment of evidence which confirms vs. conflicts with pre-existing beliefs (Birch and Bloom 2007, Woodward *et al.* 2007), and the recognition of ingroup members' faces (Bernstein *et al.* 2007). Face-perception biases towards recognizing members of one's own race, for instance, seem to develop within the first year of life (D. J. Kelly *et al.* 2007).

22 Conroy (2001).

23 The concept of infrahumanization (Castano and Giner-Sorolla 2006, Cortes *et al.* 2005) reflects research showing that people tend to think of outgroups as less human than ingroups. For example, they attribute complex emotions like resignation or admiration to themselves and members of their ingroups more than to outgroup members. There is also preliminary research suggesting that this division of people into, effectively, two classes of human being may be reflected at the level of brain function: see e.g. Harris and Fiske (2006).

24 Moral processing appears to be influenced by evidence of bodily harm (Heekeren *et al.* 2005).

25 Sade (trans. 1991), 603.

26 Ibid. 645.

27 Frith (2007), Frith and Frith (2006).

28 All quotations are taken from the unedited excerpt from Himmler's recording of the speech (Himmler 2007), available from http://www.holocaust-history.org/himmler-poznan/speech-text.shtml. Given the extent of controversy generated in some quarters by translations which refer to 'extermination', I feel obliged to give the German original of the key passage. It reads: 'Es gehört zu den Dingen, die man leicht ausspricht. "Das jüdische Volk wird ausgerottet", sagt Ihnen jeder Parteigenosse, "ganz klar, steht in unserem Programm drin, Ausschaltung der Juden, Ausrottung, machen wir, pfah!, Kleinigkeit".'

29 Klee *et al.* (1991), 163–71.

30 Quotations are from ibid., pp. xix, 4–5, 174–5, 205.

31 Lochner (1943), 11–12.

32 Quotations are from Klee *et al.* (1991), 163, 43. To argue that Nazi ideology was not *amoral*—i.e. that it recognized and valued certain moral codes—is to make no claim about whether Nazi behaviour was *immoral*, by our standards or its own.

33 For a review of disgust's role in moral psychology, see Haidt (2007), Haidt *et al.* (1997), or Miller (1997).

34 See e.g. Gu and Han (2007).

35 The *SS-Einsatzgruppen* were elite units answerable, like all the SS, to Hitler (via Heinrich Himmler and his deputy Reinhard Heydrich) rather than to the rule of law. Following the German army into Eastern Europe, their role was to enable the transition from conquered territory to lands fit for German administration and settlement. To achieve this, the *Einsatzgruppen* 'confiscated weapons and gathered incriminating documents, tracked down and arrested people the SS considered politically unreliable—and systematically murdered the occupied country's political, educational, religious and intellectual leadership' (Rhodes 2003, 4). Needless to say, their targets included Jews.

Chapter 3: Why does cruelty exist?

1 See Hobbes (ed. 1996), Rousseau (trans. 1973). The 'nasty, brutish and short' soundbite is somewhat unjust to Hobbes, who does not argue that humans are inherently nasty. His claim is that we desire peace as long as it does not conflict with our interests, but when conflicts arise which threaten our self-preservation we tend to resolve them by violence. The distinction is between an urge to hurt which is as instinctive as the urge to eat (an idea more associated with Sade), and the strategic use of harm-doing in certain situations. Hobbes's reputation, however, is probably stuck with its mythical aspects. Besides, the argument over whether vicious behaviour is 'social' or 'natural' remains, whoever's name is attached to the framing positions. Perhaps people's preferences come down to personality, or at least their

default expectations of other people's essential nature: optimistic for Rousseauvians, pessimistic for Hobbesians.

2 Attenborough *et al.* (2001).

3 See e.g. http://www.world-science.net/exclusives/050209_warfrm.htm.

4 de Waal (1996*a*, *b*).

5 Elliott (1996) discusses the importance of case-by-case assessment of moral responsibility with particular reference to mental illness.

6 Milgram's original study (Milgram 1963) instructed his participants to give what they thought were dangerous, potentially lethal electric shocks to people. To his surprise and everyone else's, up to two-thirds of his volunteers obeyed, albeit reluctantly. Experts consulted prior to the experiments had predicted a compliance rate of around 4% (Milgram 1997). The work has been widely and cross-culturally replicated.

7 See Ch. 2, n. 2; also Zimbardo (2007).

8 Pincus (2001).

9 See Dennett (1989). The psychological term 'theory of mind' refers to the extraordinarily powerful human capacity to create explanations of our behaviour and everybody else's: the conceptual framework of beliefs, desires, intentions, and so on used to interpret the activities of agents. 'Theory of mind' is a term used widely in cognitive neuroscience, especially by scientists applying for funding to research it. An alternative term, 'folk psychology', unfortunately implies that ordinary people are incapable of thinking properly and hence require an academic elite to conquer the brain and then teach them how to think better, in time removing the need for folk psychology (Churchland 1989).

10 In practice, of course, divisions are never so neat. Our default settings are so powerful that removing the language of agency from even low-level scientific explanations, let alone attempts to construe human behaviour, can be hard or impossible. T cells 'attack', proteins have 'preferred' conformations, and atoms 'seek' an electrically stable state. Even descriptions of Darwinian natural selection, that archetypally purposeless procedure, tend to use the agency framework, with references to natural selection 'designing' and 'choosing' its various outcomes. Personifying evolution is not the intention, but describing it entirely impersonally is actually quite a challenge, testifying to our addiction to agency.

11 D. J. Kelly *et al.* (2007), Foster and Young (2001).

12 Waller (2002).

13 Butler *et al.* (2007), LeDoux (1998), Luo *et al.* (2007), Nell (2006).

14 Surprise is not discussed here for reasons of space. It can be thought of as an 'early warning' mechanism, an instinctive orienting response which helps the organism identify threats quickly. Any threat response must make a trade-off

between efficiency (responding appropriately) and speed of response (Ohman *et al.* 2007). It is also worth noting here, as elsewhere, that emotions can be mixed (see e.g. Muris *et al.* 2008).

15 Dickerson and Kemeny (2004).

16 Ibid.

17 Gailliot *et al.* (2007).

18 For a study of the role of anger in anti-outgroup prejudice see DeSteno *et al.* (2004).

19 One day computers will be designed which can swear back, and then we will finally have achieved Shakespeare's rude mechanicals in a modern incarnation.

20 See e.g. Maner *et al.* (2005), van Honk and Schutter (2007).

21 Couppis and Kennedy (2008). A threat source may be physically powerful and at least potentially controllable (e.g. a social rival, triggering anger), or it may be physically powerless and yet be relatively uncontrollable (e.g. a fast-moving cockroach, triggering disgust). See Fischer and Roseman (2007).

22 Curtis *et al.* (2004), Fessler and Haley (2006), Haidt *et al.* (1994), Miller (2004). See also Kolnai (ed. 2004).

23 For a classic treatment of this topic, see Douglas (2002); also see Miller (1997) and Parker (1983).

24 Faulkner *et al.* (2004), Park *et al.* (2003).

25 Rozin *et al.* (1986).

26 Ingroup preference, also called ethnocentrism, is a universal and powerful human trait: the need for secure belonging to a clearly defined group (Sumner 1907, esp. 13–15; see also Brewer 2007). That group need not however be defined in ethnic terms (Tajfel *et al.* 1971).

27 Fehr and Gachter (2002), Fehr and Rockenbach (2004), Hauert *et al.* (2007), and Dreber *et al.* (2008) highlight recent work on altruistic punishment.

28 Dunbar (1997).

29 Altruism of any kind looks challenging for a theoretical framework with slogans like 'the survival of the fittest' and 'the selfish gene'. For such an outlook surely callousness should be the order of the day. In Europe and America numerous thinkers advocated precisely this conclusion in the late nineteenth and early twentieth centuries (Black 2004). At home, eugenics was a popular notion among elites alarmed by the restive masses (votes for women and workers' rights were high on the political agenda). Abroad, theories of racial superiority were used to defend colonial mass murder. Yet the same cultured individuals who ignored the devastation of African populations not only looked after their own children but cared for their wives' sick relatives and paid the debts of needy distant cousins. Such kindness, irrational as it

might seem on strict Darwinian grounds, has since been drawn into the evolutionary fold by William Hamilton's exquisite work on kin selection, which offered an explanation—and detailed, quantifiable predictions—of the circumstances in which altruism should and should not be observable (Hamilton 1964). Many of Hamilton's predictions have since been confirmed, and both kin selection and altruism continue to be fruitful topics of research in evolutionary biology (see e.g. Dugatkin 2006, Fehr and Rockenbach 2004).

30 Dennett (1995), cited in Dennett (2006), 67.

31 See e.g. Fehr and Gachter (2002), Johnstone and Bshary (2007).

32 Ambrose (1998), Gherman et al. (2007).

33 See Schultz et al. (1997).

34 Junior academics struggling on short-term contracts or no contract at all have been known to remark that achieving tenure has an effect on their senior colleagues analogous to that which the rock exerts upon the polyp. It's a tempting idea, but lacking in evidence. See Holley (1977), Wolfe et al. (1996).

35 Bar (2007), Holy (2007), Wolpert et al. (1998).

36 The capacity to generate expectations about, for example, the behaviour of other agents in the world appears to be one which humans possess from infancy, i.e. before language development (see e.g. Surian et al. 2007).

37 See Kolnai (ed. 2004), p. 61.

38 For more on the biological basis of the sense of control see Declerck et al. (2006), Linser and Goschke (2007).

39 Deacon (1997), Kveraga et al. (2007).

40 The error signal resulting from conflict between expectations and reality may be related to the psychological phenomenon of cognitive dissonance described by Leon Festinger (Cooper 2007, Festinger 1957), in that cognitive dissonance may be a subset of the broader phenomenon of conflict detection. If this speculation is correct, the extensive psychological literature on dissonance and the cognitive neuroscientific research on conflict monitoring could potentially benefit from a merger. For more on the roles of the cingulate cortex in conflict monitoring see Botvinick et al. (2004), Braver et al. (2001); for a review of findings on the brain's 'pain matrix' see May (2007).

41 Croyle and Cooper (1983), Elliott and Devine (1994), Hajcak and Foti (2008). The same may also apply to other species (Egan et al. 2007).

42 Stories of this kind, which present ingroup members with role models while congratulating them on being part of a group with such fine ideals, are common in science, as in other groups. Richard Dawkins recounts a similar anecdote in The God Delusion (Dawkins 2006, 283–4), and elsewhere.

43 Actions aimed at fulfilling the need for control need not be directed against the source of the challenge to one's sense of control; see e.g. Stets (1995).

44 The three areas of vulnerability and the primary emotions associated with their threat responses—existence (fear), power and status (anger), and identity (disgust)—resemble the three themes of identity, security, and purity described by Jacques Semelin (Semelin, trans. 2007).

45 The quotation is from the social scientist William Graham Sumner (Sumner 1907, 2).

Chapter 4: How do we come to act?

1 For an exhaustive description of neurons' habits, try Byrne and Roberts (2004).

2 Jones *et al.* (2007).

3 See Taylor (2004), ch. 10.

4 See e.g. Sterzer *et al.* (2002).

5 A classic study in social psychology illustrating the importance of time pressure in making people less altruistic is Darley and Batson (1973).

6 Wu and Huberman (2007).

7 The preference for ideas which fit into a pre-existing mental schema is reflected in their faster consolidation in memory (Tse *et al.* 2007).

8 Recent books addressing the theme of our limited awareness of our own brain function include Wilson (2002) and Fine (2007).

9 A classic article on vision processing is Felleman and Van Essen (1991).

10 The biblical statement is from Matthew 25: 29.

11 See e.g. Brecht and Schmitz (2008).

12 Here I draw heavily on the work of three non-neuroscientists: the philosophers Ludwig Wittgenstein, Daniel Dennett, and Peter Hacker (Wittgenstein, trans. 1974; Dennett 2003b; Hacker and Bennett 2003).

13 Dennett (2003a).

14 In slower, reflective evaluations context carries more weight than in rapid, automatic evaluations (Cunningham and Zelazo 2007). Time pressure is not the only stress. Situations where a problem is presented as requiring a definite answer create a need for closure which is less evident when ambiguity and uncertainty are acceptable, and high need for closure can produce similar effects to time pressure, such as the unconsidered rejection of alternative options and a desire for strong leadership (Pierro *et al.* 2002).

15 Buzsaki *et al.* (2007).

16 Clearly there are degrees of incompatibility in motor control. Being told to move your head forward and to the left, for instance, allows for more compromise than being told to move your head forward and backward at the same time.

17 Stern *et al.* (2007).

18 Lee *et al.* (1999).

19 Edelman (1987).

20 Indecision can delay us, but never to the fatal extent of that archetype of dithering, Buridan's Ass. This hypothetical creature began life as an Aristotelian man, briefly mentioned in the philosopher's *On the Heavens* (Aristotle, trans. 1939, II.13.iii, p. 237). The medieval version, a philosopher's donkey, was thought to remain unmoving when placed midway between food and drink because he was equally hungry and thirsty. The ass, in theory, starved to death because he could not choose between his options (possibly thus giving rise to the American phrase 'dumb ass').

21 Sparrow and Wegner (2006), Taylor (2001).

22 Friston (2005), Taylor (2001).

23 See Block (1995), Csikszentmihalyi (2002).

24 Egner *et al.* (2008).

25 Adopting the metaphor of water flow for brain input processing, one can regard familiar concepts with numerous semantic associations as providing many 'escape routes' for the flow when blockages occur. One can therefore predict that neural activity should dissipate faster for semantically rich concepts than for concepts with few associations, and there is some evidence of this (e.g. Pexman *et al.* 2007).

26 Friston (2005).

27 Capgras syndrome involves a selective failure to recognize familiar people *as familiar*. Patients may accept that the person in front of them looks like their spouse or friend, but may insist that in fact that person is a robot or impostor. Prosopagnosia involves problems with recognizing faces (but not people, who can be identified e.g. by clothes or mannerisms); while Alzheimer's patients may fail to recognize faces as part of a more widespread neurodegenerative disorder.

28 Libet *et al.* (1999).

29 Chen *et al.* (2008). Reasons not to act may come from within or from external sources (Fishbach and Trope 2005).

30 The effects of priming on attention, memory, and social cognition have been much studied; see e.g. Desimone (1996), Fahy *et al.* (1993), K. J. Jonas and Sassenberg (2006), Stone and Valentine (2005). For an example of emotional priming, see Ferguson *et al.* (2005).

31 Fogassi *et al.* (2005), Haslinger *et al.* (2005), Jeannerod and Frak (1999), Pfurtscheller *et al.* (1999).

32 Browning (1991).

33 Martens *et al.* (2007).

Chapter 5: How do we come to feel?

1 Hillenbrand and van Hemmen (2002), Miyata (2007).

2 See Edwards (1988), Hornby (2001).

3 Stress, for example when doing a difficult task, has a deleterious effect on mental flexibility. Noradrenaline is thought to be involved in mediating this deterioration (Campbell *et al.* 2008).

4 René Descartes, *Discourse on the Method* (in Descartes, trans. 1988, see pp. viii, 36).

5 The intellectual face of logicalism was logical positivism, which ranked statements depending on whether they could be verified as true or false. Key texts include Ludwig Wittgenstein's early masterpiece the *Tractatus Logico-Philosophicus* (Wittgenstein, trans. 2001) and the *Principia Mathematica* by Alfred North Whitehead and Bertrand Russell (Russell and Whitehead 1910–13). Logicalism, a much broader term, refers to the tendency to value (in some cases to the point of worship) logic, mathematics, and positivist reasoning. It can be found in the scientific justifications offered for racism, eugenics, colonial brutalities, and unethical government experiments (see Ch. 8, n. 9). It is also prominent in modern scientism and scientific atheism.

6 Wittgenstein, for instance, set aside philosophy for some years after publishing his *Tractatus*. His later work, such as the posthumously edited and published *Philosophical Investigations*, took the very different approach which made him one of the most influential thinkers of the twentieth century (trans. Wittgenstein 1974). For details, try the *Investigations*, or see Ray Monk's biography of Wittgenstein (Monk 1990).

7 For guidance to the debates about the unity/diversity of emotions, see Adolphs *et al.* (2003), Barrett, Mesquita, *et al.* (2007), Hennenlotter and Schroeder (2006), Scherer and Wallbott (1994). Some of the conceptual hazards of affective neuroscience are highlighted by Peper (2006).

8 Darwin (1999); Ekman *et al.* (1969).

9 The body's somatosensory systems are concerned with information about feelings of touch, texture, pressure, body position, and the like. They are normally thought of as 'sensory', as opposed to 'visceral' (i.e. processing information about

one's squamous interior), as the two appear to be distinct. For example, visceral information appears to travel, in part, through sensory nerves which are not surrounded by the myelin sheath which makes most nerves extremely fast signal carriers. Thus visceral information takes longer to reach the brain than somatosensory input (Castell *et al.* 1990; see also Aziz *et al.* 2000).

10 Nordgren *et al.* (2007).

11 Believe it or not, anal and rectal extension studies are a valued part of modern neuroscience (e.g. Eickhoff *et al.* 2006), and can even be done in an fMRI scanner, with ingenuity. (The ethics committee application for that project must have been a peach.)

12 The somatic marker hypothesis is discussed in Bechara *et al.* (2000; 2005), Damasio *et al.* (2000). Damasio also presents his views on emotion processing in the brain in three books (Damasio 1996; 2000; 2003).

13 Needless to say, I simplify, as correlation is not the only mechanism through which children acquire language. For more details, see Barrett, Lindquist, *et al.* (2007), Yu and Smith (2007).

14 Papafragou *et al.* (2007).

15 Care-giver teaching may also occur through less explicit mechanisms. Research suggests, for example, that parents' sensitivity to disgust affects their offspring's chances of developing phobias to particular animals (Davey *et al.* 1993).

16 See e.g. M. D. Lieberman *et al.* (2007); also Schachter and Singer (1962).

17 Gregory *et al.* (2003). Written emotion words, for instance, are remembered better than neutral words, and appear to be treated differently by the brain even early on in processing (Kissler *et al.* 2007).

18 Clore and Huntsinger (2007).

19 Damasio's somatic marker hypothesis has been hugely influential. For a recent critique see Dunn *et al.* (2006).

20 Schachter and Singer (1962).

21 Davey (1994), Davey *et al.* (1998; 2003).

22 Phillips *et al.* (1998).

23 Disgust may also trigger changes in skin conductance and reductions in heart rate and blood pressure (Rozin and Fallon 1987, Stark *et al.* 2005) as well as breathing changes due to increased upper airway resistance (Boiten 1996, Ritz *et al.* 2005), changes which appear to be specific to disgust (Collet *et al.* 1997). Signals of gastrointestinal distress (e.g. feelings of 'churning' or queasiness, nausea, gagging, and vomiting) and nonverbal vocalizations, withdrawal, avoidance, and self-cleaning reactions may also occur (Curtis *et al.* 2004). EEG (electroencephalography) studies, which measure the electrical 'brainwaves' emitted by active neurons, show that disgust

evokes distinct patterns of cortical activity relative to other emotions, both with respect to spatial distribution (Aftanas *et al.* 2006, Sarlo *et al.* 2005) and time course (Esslen *et al.* 2004).

24 Kuniecki *et al.* (2003), Levenson *et al.* (1990).

25 Hornby (2001), Saito *et al.* (2003).

26 The area postrema may also be activated directly via the bloodstream.

27 Researchers have proposed a 'sequential activation' model in which a thresh-old level of input to the brainstem control centres must be reached before vomiting can occur (Edwards 1988, Hornby 2001). Input signals must accumulate, and may trigger associated symptoms like nausea before they accumulate sufficiently to set off the full response. This prevents vomiting, which uses a lot of energy, being provoked unless there is a real need to throw up.

28 The NTS connects to the area postrema (and vice versa); it is also intercon-nected with multiple other areas including the hypothalamic paraventricular nuclei, thalamus, amygdala, somatosensory and motor cortex, and perirhinal and insular cortex. See Buller (2003), Landis *et al.* (2006), Saha (2005), Sequeira *et al.* (2000), Sewards (2004).

29 Broussard and Altschuler (2000), Hornby (2001).

30 Travagli *et al.* (2003), Travagli and Rogers (2001), Cameron (2001).

31 The somatosensory cortex and particularly the insula process information about taste, pain, and the state of the body's visceral organs. Studies which have recorded electrical signals from neurons in the insula (research done in human patients undergoing brain surgery, e.g. for severe epilepsy) show that these neurons fire in response to images of other people's disgusted expressions—but not when presented with faces showing other emotions (Krolak-Salmon *et al.* 2003). The insula also receives signals from the vestibular organs in the inner ear, which are thought to trigger motion sickness. Neuroimaging studies which look at how brains react to disgusting stimuli show insular responses to revolting images, movies, smells, tastes, texts, and even sounds (e.g. Phillips *et al.* 2004, Wicker *et al.* 2003). Finally, stimulation of the insula in humans alters blood pressure and heart function; it can also trigger disgusted expressions and unpleasant sensations in the mouth, face, and gut (Krolak-Salmon *et al.* 2003, Naidich *et al.* 2004). Electrical stimulation of the anterior cingulate likewise produces autonomic changes (e.g. in heart rate, blood pressure, and breathing), as well as visceral responses, including nausea and vomiting (Benarroch 1997).

32 Harris *et al.* (2008), Kramer *et al.* (2007), Sterzer *et al.* (2007).

33 For an example of emotional habituation to media violence see C. Kelly *et al.* (2007).

34 For 'trolley' some readers may prefer to substitute 'train' or 'tram'. For a description of the trolley problem see Greene *et al.* (2001). See also Waldmann and Dieterich (2007).

35 The original 5-or-1 trolley problem was developed by the philosopher Philippa Foot (Foot 1978), in an article first published in 1967. Her version had a runaway tram and workmen, rather than ramblers. This does not materially affect my argument.

36 Hassin *et al.* (2007).

37 For an example of emotion-congruence effects involving disgust, see Davey *et al.* (2006). See also Forgas (1998), E. Jonas *et al.* (2006). For an example of the use of emotion words in propaganda, see Taylor (2004), 151.

38 Dennett (1989).

39 For *persistent* child cruelty there is evidence of a link with later conduct disorder, violence, and psychopathy (Hensley and Tallichet 2005*a*, *b*; 2008; see also Merz-Perez and Heide 2003).

40 For more on the relationship between disgust and moral judgement see Trafimow *et al.* (2005), Wheatley and Haidt (2005), Zhong and Liljenquist (2006), or for an overview, Jones (2007).

41 As stated earlier (for sample references see Ch. 1, n. 23), there is some evidence that this effect may extend to the courtroom, with juries showing more severity towards unattractive criminals. Research suggests that people particularly associate ugliness with more violent crimes and also with mental illness. See e.g. McKelvie and Coley (1993). This has been a controversial area of research at least since the now-discredited work of Cesare Lombroso linked physical ugliness to criminality. Whether such a link actually exists is of course a separate issue from whether the perception of a link is widespread, and if so, what impact that has upon moral judgements.

Chapter 6: How do we come to believe?

1 Austin *et al.* (2006).

2 The existence of non-synaptic information transmission has in fact been known for some time (see e.g. Jourdain *et al.* 2007, Vizi and Mike 2006).

3 For a glimpse of the alarming complexities surrounding synaptic change, see the reviews by Lynch *et al.* (2007), Massey and Bashir (2007), Raymond (2007), or Thiagarajan *et al.* (2007).

4 Hinton (2002).

5 Plaks *et al.* (2005).

6 For an introduction to the psychology of unconscious processing, see Wyer (1997).

7 Kveraga *et al.* (2007).

8 A classic text on this aspect of group dynamics, which gave us the term 'group-think', is Janis (1982).

9 Challenges successfully seen off can make the belief more strongly held (Tormala and Petty 2002).

10 For a discussion and critique of identity politics see Barry (2001).

11 Barry (2001).

12 The influential terror management theory of Greenberg and colleagues (Greenberg *et al.* 1997) has inspired numerous studies of the relationship between intimations of mortality and social belonging. See e.g. Arndt *et al.* (2002), Florian *et al.* (2002). In brief, terror management theory suggests that numerous human behaviours can be explained as defensive reactions to 'mortality salience', the awareness that one will die.

13 The meme analogy derives from Richard Dawkins's comparison of cultural and genetic transmission, in which he postulated that, just as genes carry information between generations, so memes can carry information between brains. Dawkins has also compared memes, particularly memes of which he doesn't approve, to viruses. See Blackmore (2000), Dawkins (1989).

14 An influential paper arguing that cultural evolution does not need memes is Henrich and Boyd (2002). See also Rogers and Ehrlich (2008).

15 See Laqueur (2004), 175.

16 The first two quotations are from Hitler (trans. 1969), 54, and a speech made by Mao Zedong to the Chinese Communist Supreme State Conference on 27 February 1957 (and published on 19 June that year in the *People's Daily*). The 'smallpox' quotation is from Richard Dawkins (Dawkins 1989, 330). See also John Cornwell's rebuke of Dawkins for describing religion as 'a bacillus' (Cornwell 2003, 137–45, and Dawkins 2006, esp. 186–8).

17 Re-education, incidentally, is the way Chinese Communists described their methods of making US prisoners of war in Korea (1950–3) recant and denounce their government. The government in question, well aware of the importance of propaganda, called its enemy's techniques brainwashing and portrayed them as the latest Red menace (Taylor 2004).

18 Cathedrals as concentration camps, perhaps? If that idea strikes you as shocking, remember that churches have become mass graves in many atrocities (as I write, the news is full of just such a massacre, in Kenya).

19 See e.g. Bushman *et al.* (2007), Loza (2007).

20 Hodson and Costello (2007), Taylor (2007).

21 The role of 'brainwashing' techniques in wartime atrocities is discussed in Taylor (2006).

22 Klee *et al.* (1991), 259.

23 The comment 'this is not the whole story', which might serve as a motto for this entire book (and indeed for any scientific discussion), is particularly applicable here. Of necessity I am glossing over a huge literature, ranging from research on conflict monitoring in the brain (see e.g. Botvinick *et al.* 2004, Egner and Hirsch 2005), through cellular neuroscientific studies of how synapses change (see Ch. 6, n. 3), to work on neural synchronization and the role of oscillations (see e.g. Buzsaki 2006), all of which is likely to be crucial to understanding how neurons resolve their differences.

24 Case *et al.* (2006).

25 Langer (1999), 2.

26 Green (2007), 192–211, Danner (2005), 52. The massacre at El Mozote was part of *La Limpieza* (see Ch. 2, n. 18).

27 Research suggests that there is a psychological association between physical and moral purity, such that enhancing physical cleanliness may make one feel morally cleansed (Zhong and Liljenquist 2006). Nazi ideology was only one of many belief systems to blur the distinction, promoting physical health as if it would automatically lead to moral well-being.

28 Hitler (trans. 1969), 226, 396.

29 The quotation on education is from *Mein Kampf* (ibid. 389; see also Klee *et al.* 1991, p. xiv; Lifton 2000, first section). The euthanasia programme was named after 'Tiergartenstrasse 4', the address of its Berlin headquarters.

30 Lifton (2000).

31 The quotation from Fritz Klein is cited in ibid. 16.

32 Hilberg (1985), 18; see also Lifton (2000), 16.

33 Klee *et al.* (1991), 217.

34 For the context of the 'do no harm' sentiment, see Hippocrates' *Epidemics*, I.xi (Hippocrates, trans. 1923).

Chapter 7: Why are we callous?

1 Card (2002).

2 Machiavelli's *The Prince* argued that 'it is far better to be feared than loved if you cannot be both' (Machiavelli, trans. 1961, 96), but that cruelty should be used instrumentally (i.e. callously). A prince commanding an army, for instance, needs a reputation for cruelty, because without it 'he can never keep his army

united and disciplined' (p. 97). Machiavelli cites the Carthaginian general Hanni-bal, arguing that only his 'inhuman cruelty' kept his huge and disparate forces together.

3 The modern rules of battle conduct can be found in the two bodies of the law of war, the *ius ad bellum* (when to pick a fight) and the *ius in bello* (how to pick a fight). See Ch. 1, n. 8.

4 Useful starting-points in game theory can be found in works by one of its great practitioners, Robert Axelrod (Axelrod 1984; 2003).

5 See e.g. Burris and Rempel (2004). The 'ocean of notions' is brought to life in Salman Rushdie's *Haroun and the Sea of Stories* (Rushdie 1991).

6 See Rejali (2008) on how the pressure for secrecy, given disapproving public opinion, has led democracies to develop particularly stealthy methods of torture.

7 D. Lieberman *et al.* (2007), Platek *et al.* (2008).

8 There is evidence that stepchildren are at higher risk of domestic abuse than biological offspring (Daly and Wilson 1998). More cross-cultural research is needed on this topic, since cultures vary in the importance they place on biological parentage. See also Buss (2005) and Kurst-Swanger and Petcosky (2003) on domestic violence.

9 See e.g. Jenkins *et al.* (2008). Psychological similarity between two individuals may lead to mutual appreciation, but there are of course caveats. For example, two individuals may be extremely similar and yet detest each other, because they differ or have come to differ on one issue which is important to both of them. When making comparisons of similarity, there is also the question of whether the aspect of the self involved is ideal or actual; people who share some unwanted feature of the self may be liked less.

10 Both physical and symbolic kin groups, of course, can be seething cauldrons of conflict, especially if the main threats they face are not human attackers but limited resources. However, much of the cruelty which occurs in such cases is either relatively minor (e.g. verbal abuse) or an impulsive and short-term response to some perceived threat. Otherization can sometimes become severe, leading to escalating cruelty. Yet since otherization involves participants deliberately differentiating them-selves from their opponents, the symbolic kinship of perceived similarity is likely to disintegrate as otherization worsens.

11 The use of the terms 'thick' and 'thin' with reference to cultural signs is asso-ciated with the anthropologist Clifford Geertz, who adopted it from the philosopher Gilbert Ryle. Ryle used the terms to refer to descriptions of actions: for example, the same wink could be described thinly as a muscle contraction or thickly as a conspiratorial signal. Geertz extended the usage to descriptions of culture. See Geertz (1973), Ryle (1971*a*, *b*), or for a similar, more recent approach to action descriptions, see Wegner (2002).

12 For references to Hamilton's work see Ch. 3, n. 29.

13 Brogden (2001), Buss (2005), Hausfater and Hrdy (1984).

14 When a conflict between two similar people becomes apparent to one or both, the difference between them will be highly salient because of its distinctiveness. In addition, the associated negative emotions will tend to make other, minor differences more salient, if these are also linked to negative emotions, due to the phenomenon of emotion congruence (discussed in Ch. 5).

15 Navarrete and Fessler (2006), Stevenson and Repacholi (2005).

16 The use of the lazar-house as a form of social control is discussed in Moore (1987).

17 Delegating nursing care to older women (i.e. placing the risk on those whose direct reproductive contributions to the group have already been made) is an efficient way of boosting sick individuals' chances of survival while minimizing the dangers of damage to the group's reproductive potential.

18 Kruglanski et al. (2002).

19 Blair (2005), Shamay-Tsoory and Aharon-Peretz (2007).

20 Gallese et al. (1996), Leslie et al. (2004), Rizzolatti (2005).

21 See de Vignemont and Singer (2006), Singer et al. (2004; 2006). It is worth noting, with respect to MRI studies, that just because two different processes (e.g. experiencing pain and pain empathy) activate the same area of the brain does not mean either that identical sets of neurons are involved or that the brain areas are being used in the same way (see Morrison and Downing 2007).

22 Social neuroscience research suggests that autism primarily involves problems with cognitive empathy, while deficits in emotional empathy are found in psychopaths. People with autism can and do experience strong negative emotions, such as fear. People with psychopathy understand that fear affects other people, but seem to feel it much less intensely themselves (Blair et al. 1997). Empathy for others' pain can also be reduced if the person empathizing is in pain as well (Valeriani et al. 2008).

23 This principle, retroactively applied, is the basis of programmes of restorative justice, which aim to make offenders feel, not just understand intellectually, the damage they have done. Restorative justice, which has recently become more widely used in Britain, forms the basis for many other judicial systems, and seems to be effective in many cases. More information can be found at the Restorative Justice Consortium (2007) or in a recent review (Sherman and Strang 2007).

24 Klee et al. (1991), 142.

25 Ibid. 154. Häfner's statement was made in 1965; in 1973 he was sentenced to eight years in prison.

26 Arendt (1963), cited in Milgram (1997), 23.

27 Quotations are from Klee *et al.* (1991), 143–51.

28 Ibid. 151.

29 Babi Yar, in the mountains near Kiev, contained a ravine which, an observer of the massacre reported, 'was about 150 metres long, 30 metres wide and a good 15 metres deep'. Accessible only through narrow entrances, the execution zone could thus be kept separate from the waiting area where the Jews were stripped of their clothes and possessions. On 29/30 September 1941, 33,771 men, women and children were shepherded into the ravine, made to lie down, and shot (ibid. 64–8). The ravine was then partially blown up in order to bury the corpses.

30 Quotations are from ibid. 96–7. Italics are as in the original.

31 Quotations are from ibid. 95, 89.

32 Rees (2005).

33 For references on the Rwandan genocide see the Introduction, n. 23. The mass killing in the Polish village of Jedwabne in 1941, a stark case of neighbour turning on Jewish neighbour, is detailed in Gross (2003). The atrocity committed by American troops against the villagers of My Lai in 1968, during the Vietnam war, was brought to public attention by Seymour Hersh (Hersh 1972). For a guide to the story of Hiroshima and Nagasaki, see Rhodes (1988).

34 Widespread in medieval Europe, the blood-libel was the claim that Christian infants were snatched and sacrificed by Jews so that their blood could be used ritualistically in the observances of Passover (see Frankfurter 2006, esp. 149). The story of Djordje Martinovic and the painfully inserted bottle, which may or may not have been inflicted by malevolent Kosovans, transfixed the Serbian press in 1985. It is briefly recounted by Noel Malcolm (Malcolm 1998, 338), and given a more detailed treatment by Julie Mertus (Mertus 1999, ch. 2).

35 Frankfurter (2006), 12.

36 Tilly (2003), Valentino (2004).

37 Even so apparently abstract a feeling as feeling safe has its neural correlate these days; in this case the ventrolateral prefrontal cortex (Bender *et al.* 2007).

38 Prunier (1995), 171–2, cited in Hinton (2002), 159. See also Gourevitch (2000), 96.

39 Atran *et al.* (2007), Ginges *et al.* (2007). Fascinating discussions of the concept of 'the sacred' can be found in Parker (1983) and Girard (trans. 2005).

40 The historian of terrorism Walter Laqueur (Laqueur 2004) traces the roots of jihadist thinking in Islam to Ibn Taymiyyah (b. 1268). His views were considered controversial and unofficial by his contemporaries, and the reinterpretations of modern 'radical Islam' by Sayyid Abu ala Maududi, Sayyid Qutb, Osama bin Laden, Ayman al-Zawahiri, and others are also challenged by other Muslims. They are not the pure

return to founding principles that the term 'fundamentalism' implies. See also Burke (2004), Ruthven (2004).

41 Schachter and Singer (1962).

42 Chirot and McCauley (2006).

43 For an examination of the role of paranoia in politics, see Robins and Post (1997).

44 Dallaire (2004), 255.

45 Frankfurter (2006).

46 The fourteenth-century Battle of Kosovo, which pitted Serbia against the Ottoman Empire, remains a prominent part of Serbian national mythology (Malcolm 1996; 1998; see also Mertus 1999). For more on the role of essentialism in large-scale intergroup atrocities see Weitz (2003).

47 Preston (1994).

48 Chirot and McCauley (2006), 36.

49 The *Daily Mail* was not alone in making the British climate of the time inhospitable to Jews. The editor of *The Times*, for example, called for all Jews to be removed from official service (Pugh 2006, 215). Both newspapers, Martin Pugh argues, reflected the widespread anti-Semitism of the time.

50 Many expressions of racism never reach the justice system. Of those which do, UK Home Office figures for 2006/7 (http://www.homeoffice.gov.uk/rds/crimeew0607.html) suggest that racially or religiously aggravated crimes comprised 2% of assaults without injury and 11% of harassment offences (Nicholas *et al.* 2007).

51 Fein (1990).

52 Marcel Duchamp's well-known *Fountain*, an artwork consisting of a urinal, plays on this ambiguity of symbol and function. As far as I know, however, the object itself was clean. Even more visceral is the work of artist Piero Manzoni, who in 1961 tinned his own excrement, ninety times.

53 See, for example, my discussion of this point in Taylor (2004) and (2007). Nazi propaganda is notorious for associating Jews with rats and cancer, referring to 'the Jewish bacillus', etc.; but many other perpetrators have used similar terminology. Hard-line officers in El Salvador described Communism as a cancer (Danner 2005, 49). Perpetrators of the Rwandan genocide referred to their victims as 'inyenzi' (cockroaches; see Dallaire 2004, 142). Japanese soldiers who made use of 'comfort women' (women forced into hideously degrading prostitution, in cultures where female chastity was highly prized) during the Second World War referred to them as public toilets (Chang 1998, 53).

54 Parker (1983).

55 The Nazis' Madagascan plan was later deemed infeasible (Rees 2005).

56 More information about Babi Yar can be found in Ch. 7, n. 29. Hinton (2002) has much to say about the ways in which local culture mediates mass killing, using the example of Khmer Rouge atrocities. Christopher Taylor's chapter considers the anthropology of the Rwandan genocide.

Chapter 8: Why does sadism exist?

1 Psychoanalytic or queer theory approaches to sadism will not be mentioned here, for lack of space. Life and this book's word limit are simply too short to do those literatures justice. Discussing sadism as a form of cruelty also excludes voluntary participation as part of sexual practice, since any pain inflicted on a willing victim is not suffering in the sense used here.

2 For an indication of the complexities involved, see Gray *et al.* (2003). The term 'personality disorder' is particularly problematic because of the essence trap: the difficulties raised by describing as pathological not particular thoughts and feelings (as in major depression, for example), nor even behaviours (as in kleptomania or voyeurism), but an individual's entire character.

3 The paucity of research on sadism is not simply because of the nature of the topic; there is plenty of scientific interest in psychopaths, a clinical population which overlaps with that of sadists (Porter *et al.* 2003). It may have more to do with the dominance of psychoanalytic perspectives in the study of sadism to date and with the moral aura surrounding it. Psychopathy is considered a more respectably scientific concept.

4 See for example the descriptions of sexual sadism and sadistic personality disorder in the psychiatric 'bible', the DSM (American Psychiatric Association 2000). The clinical (as opposed to literary) association of cruelty with lust owes much to Krafft-Ebing (see Ch. 1, n. 2).

5 Browning (1991), Zimbardo (2007).

6 A discussion of prevalence estimates for sadism and the difficulty of obtaining them can be found in Porter *et al.* (2003).

7 Kinsey *et al.* (1953). For a psychoanalytic/critical-theoretic perspective on the relationship between sadism and masochism, see Deleuze (trans. 1971).

8 Videos of suicide bombers can present the perpetrators as brave and manly heroes. This is simply false. If, as some claim, they truly believe they will be rewarded after death then the oblivion granted by triggering a suicide vest is nothing more than a coward's easy shortcut: instant gratification for the faithful. If not (and not all suicide terrorism is religiously motivated), this method of killing oneself is still surer and less unpleasant than many. An elderly woman slowly dying of cancer

needs far more courage than any suicide bomber. Suicide bombers may be idealistic, despairing, furious, or coldly determined. They may genuinely long to help the others whose suffering they observe. They can even be intelligent and well-educated individuals (Gambetta 2005), as can torturers and those who sanction them. Yet using either tactic is arguably stupid, given the lack of evidence that either is effective. Worse, however, is the tendency of both torturers and terrorists to pretend that their behaviour is just another branch of the noble art of war, extolling those 'brave' enough to do the dirty work—especially since all too often that work involves the immoral pursuit of conflict for personal advantage. If there is a better illustration of moral imbecility than this inversion of common sense I have yet to find it.

9 Before we all celebrate our escape from primitive barbarism into the clear light of secular scientific modernity, two caveats are worth bearing in mind. First, scientists and doctors are not immune to the need for control and do not always adhere to professional standards. The Second World War is known for the atrocities carried out by Nazi and Japanese doctors. Less well known are the harmful, sometimes lethal procedures on ill-informed and vulnerable people carried out in the post-war West (examples include tests of the neurotoxin sarin by the British government at Porton Down, CIA experiments, the Tuskegee syphilis trial, and the innumerable contributions made by scientists and doctors to the crafting of weapons of war and torture: see e.g. Rejali 2008). Secondly, someone labelled 'sadistic' by the power of modern Western medicine is automatically considered dangerous. If that person cannot be cured he or she must be removed from society for society's protection. This may involve long confinement or execution—not so very different from the punishment, death, or expulsion meted out by 'primitive' groups using basic morality.

10 Nisbett and Cohen (1996).

11 I say 'may' because I am not aware of research definitively identifying the causal impact of *sadistic* treatment in childhood on later parenting. There is considerable research on the effects of childhood experiences of *violence*; see e.g. Arseneault *et al.* (2006), Ballif-Spanvill *et al.* (2003), Douglas (2006), Gershoff (2002), Koenen *et al.* (2003), Ng-Mak *et al.* (2004).

12 A problem with arguments from evolution such as those presented in this book is that they are not directly testable, owing to the current inability of physicists to present us with a viable mode of time travel. Indeed, it has been said that evolution can explain anything, and thus explains nothing. The arguments presented in *Cruelty*, however, depend on certain assumptions which could be challenged empirically. For instance, the depiction of early humans as living in small groups which had to compete for resources could be undermined by future archaeological or anthropological evidence, comparisons with similar species, such as chimpanzees, or cross-species work on the effects of small-group living and resource restriction on behaviour. Evolutionary hypotheses can also be tested indirectly, by testing the predictions to which

they give rise. For example, *Cruelty*'s proposal that sadism emerges from callousness when callousness is associated with intrinsic rewards is in principle testable using psychological methods and neuroimaging, although in practice ethical objections make this difficult research to carry out. (One could also look for instances of sadistic behaviour without prior callousness, which would disprove the proposal.) Field testing in warfare, which would have greater ecological validity than laboratory-based research with respect to testing these predictions, is currently impractical, not least because MRI scanners are highly sensitive to movement and very far from portable (other technologies, such as EEG, do not offer such good spatial resolution). Research on simulated warfare, however, is under way; see e.g. Salminen and Ravaja (2008).

13 The logic of efficiency is mechanism-independent, applying to cultural evolution as well as to genetic transmission.

14 Dennett (2006), 57.

15 Jones and Fabian (2006), Keeley (1996), McCall and Shields (2008).

16 Examples of shocking cruelty—and shocked reactions to it—can be found, for example, in Euripides' *The Bacchae*, in which King Pentheus is dismembered by a group of women, including his mother, and in the legend of Tereus (told by Ovid in his *Metamorphoses*), who raped, mutilated, and imprisoned his wife's sister, Philomela. When Tereus' wife learned of this she took revenge by killing their son, Itys, and serving him up as dinner to his father (Euripides, trans. 1973, Hughes 1997, 229–45).

17 Research into homicidal fantasies suggests that '91 percent of men and 84 percent of women have had at least one such vivid fantasy about killing someone' (Buss 2005, 8).

18 Brain damage and antisocial behaviour have been firmly linked at least since Phineas Gage acquired neurology's most famous hole in the head when an iron rod destroyed part of his prefrontal cortex. The case, in which Gage appeared to change from respectable citizen to moral imbecile, is discussed in many neuroscience books, e.g. Damasio (1996).

19 My remarks about the developmental impact of sadism are speculative, since the effects of childhood physical, sexual, or verbal abuse (not all of which, of course, is sadistic) on adults are not fully understood. Not everyone shows signs of damage, but some are very severely traumatized. Most do not go on to be abusers, but some do. See e.g. Browne and Finkelhor (1986), DiLillo and Damashek (2003), Ertem *et al.* (2000), Finkelhor (1990).

20 Nutritional deficiencies are thought to contribute to aggressive behaviour (Gesch *et al.* 2002).

21 See e.g. Kirsch and Becker (2007), Marshall and Kennedy (2003).

22 Semelin (trans. 2007), Valentino (2004).

23 Foxe (ed. c.1910).

24 Nietzsche, like Sade, doesn't hesitate to attack Christianity, but where Sade emphasizes its attempts to constrain human cruelty and aggression, Nietzsche accuses it of exhibiting the urge to dominate. 'A certain sense of cruelty towards oneself and others is Christian; hatred of those who think differently; the will to persecute' (Nietzsche, trans. 1968, 131). Both writers agree, however, that Christianity undermines human potency and freedom. 'Hatred of *mind*, of pride, courage, free-dom, *libertinage* of mind is Christian; hatred of the *senses*, of the joy of the senses, of joy in general is Christian . . .' (ibid.; italics are as in the original). See also Miller (1990).

25 The story of Prometheus can be found in Aeschylus (trans. 1922).

26 For a detailed historical examination of the pleasures of warfare, see Bourke (2000).

27 The discussion of Hannibal Lecter reflects his portrayal in the books and films *Hannibal* and *The Silence of the Lambs*, as it is a very long time since I read *Manhunter*. I have not seen the latest chapter in the saga, *Hannibal Rising*, in which an attempt is made to explain the formation of Lecter's character; it did not receive the acclaim of its predecessors. With someone as representative of evil as Hannibal Lecter, shedding too much light on the darkness runs the risk of destroying the character's appeal.

28 As we shall see in Chapter 9, even when victims do not survive there is mean-ing to be found in death sadistically inflicted, demanded by some agent's ferocious desires (see Dutton 2007, 123–9). Death, particularly sudden death, can rip away hugely important segments of self in those bereaved, leaving them in shock, much as a physical wound would. To cope with that monstrosity it helps to have a perpetrator to blame, so that some sense of worth and purpose can be salvaged from the ruin of a life. The same logic, incidentally, applies if the agent is interpreted as good, rather than evil, and the suffering seen as somehow pedagogical—the point then being that God cares enough to teach us.

29 Wyndham (1960), 98.

30 Berg (trans. 1938).

31 For details of Kürten's childhood, see Berg (trans. 1938), ch. 3.

32 For more on the political and communicative aspects of terrorism, with specific regard to suicide terror, see Bloom (2005).

33 See e.g. Adam Curtis's TV documentary *The Power of Nightmares* (Curtis 2004).

34 Laqueur (2004), 403.

35 Burns (ed. 1993). Italics are as in the original.

36 Pavlov (trans. 1941).

37 Hertel and Donahue (1995), De Roos *et al.* (2004).

38 e.g. Peter Kürten (Berg, trans. 1938, 111).

39 Frith *et al.* (2004) should tell the interested reader everything he or she wants to know about bowerbirds.

40 It is possible that sexual selection by females may also favour males who are tolerant of disgust. Since women and their infants are especially vulnerable to infectious diseases during pregnancy and early postnatal life, a male prepared to deal with disgusting objects could be useful (Fessler 2002, Fessler *et al.* 2005). Demonstrating this capacity, through what I have elsewhere called 'disgust ordeals' (Taylor 2007), would therefore serve as a marker of male fitness. This may be part of the evolutionary reason why perpetrators both render their victims disgusting (e.g. through mutilation, applying excrement, etc.) and behave as if tolerating disgust has become a mark of status, e.g. posing with a corpse (as was done by US personnel at Iraq's Abu Ghraib prison following the 2003 invasion of Iraq).

41 The story of Las Poquianchis was made into a film (directed by Felipe Cazals) and a novel by Jorge Ibargüengoitia (Ibargüengoitia, trans. 1983).

42 Richardson and Hammock (2007). An example of the gender difference in public attitudes to murderers is the British media's treatment of the Moors Murderers Myra Hindley and Ian Brady. Brady, the controlling senior partner in their relationship, has not attracted anything like the levels of loathing directed at his girlfriend.

43 Cruelty by women also makes more sense if it is a display of power, since this implies that it will emerge when women are in social roles where they are expected to adopt male-type behaviour and to act callously.

44 This inclination has been called 'Herostratos syndrome', after the man who burned down a famous temple in order to gain some form of immortality (Borowitz 2005).

45 Gambetta (2005) is an authoritative introduction to research on suicide terrorism.

46 See Leknes and Tracey (2008). Pain can also be used to relieve psychological distress, e.g. in adolescent self-harm (Whitlock *et al.* 2006).

47 Blair (2005).

48 If sadism can be considered an addiction, a distortion of normal reward processing in which fulfilling the need for control becomes an end in itself, we should expect to see similarities between sadists and other addicts: similar escalating patterns of behaviour, the co-occurrence of multiple addictions, a tendency to relapse triggered by situational cues even after long periods of abstinence, the need for increasingly intense stimulation to achieve the same reward, and so on. We should also expect to see brain changes, including altered synaptic plasticity and abnormalities in dopamine metabolism and changes in reward-related areas such as the nucleus

accumbens and prefrontal cortex. (See Hyman *et al.* 2006, Kauer and Malenka 2007, for reviews of the neural structures and mechanisms involved in addiction.) These are testable hypotheses, although to my knowledge they have not to date been tested.

49 Nell (2006).

50 Pincus (2001).

51 Cross and Matheson (2006).

52 Balakian (1998), cited in Waller (2002), 52.

53 Nailing horseshoes to victims' feet (feasible lengthwise if not necessarily width-wise) was allegedly the signature torture of one high-ranking Turkish commander, Djevdet Bey. The 'horseshoer of Bashkalé' had a ferocious reputation, as Henry Mor-genthau, US ambassador to Turkey at the time, notes in his report of the Armenian tragedy (Morgenthau, ed. 2000, e.g. 204–5). Whether Djevdet Bey was responsible for this particular atrocity, and why he selected that particular form of torture, I do not know. (The horseshoe is traditionally a protection against witches and the evil eye.)

54 Zimbardo (2007), 7.

55 Browning estimates the proportions of killers who refused or evaded their murderous duties as less than 20%, noting that this is similar to findings from the much smaller sample tested in the Stanford Prison Experiment (Browning 1991, Zimbardo 2007).

Chapter 9: Can we stop being cruel?

1 For more on the role of bystanders in genocide, see Power (2003).

2 For a detailed discussion of factors influencing the relationships between vio-lence and non-violence see Barak (2003).

3 Alternatives such as physical punishment could provide the requisite emotional intensity, but cannot solve the problems of ineffective, lengthy, and downright unjust legal processes. They might also deter the public from informing on friends and family and perhaps legitimize cruel behaviour by individuals. Furthermore, state-authorized violence is dangerously prone to spreading (Rejali 2008). In short, there are numerous reasons why physical punishment is not the solution to the problem of criminal behaviour.

4 Cohen *et al.* (2007).

5 For more on temporal discounting see Ainslie (2001), Berns *et al.* (2007).

6 Shema *et al.* (2007).

7 The source of this adulation for *le divin marquis*, 'cet esprit le plus libre qui ait encore existé', was the French poet Guillaume Apollinaire (Apollinaire, ed. 1964, 194).

8 Sade (trans. 1991), 602–3.

9 Ibid. 491, 494, 518–19.

10 Quotations are from ibid. 607–8, 514, 497.

11 For examples, see Dawkins (2006) and Hauser (2006).

12 See e.g. Slote (1990).

13 Hume (ed. 1975), see esp. 34.

14 Here again I must acknowledge my debt to Wittgenstein's *Philosophical Investigations* (Wittgenstein, trans. 1974).

15 Blair made this statement in a speech on terrorism shortly after the London bombings of 7 July 2005 (BBC, 2007).

16 Further details of this view of free will can be found in chapter 11 of Taylor (2004). See also Dennett (2003b).

17 More precisely, whether you are in charge is something you can monitor and which affects your behaviour (Metcalfe and Greene 2007).

18 Foster and Young (2001) argue that even for entirely rational agents, predicting behaviour in some circumstances is impossible.

19 Wilkinson (2005).

20 Klee *et al.* (1991), 81–2.

21 See LaCapra (1997) for a detailed discussion of the phrase 'Here there is no why'.

22 If in doubt, you may wish to try a quick Internet search on the Armenian genocide. By simply using that phrase I may have condemned this book to oblivion in Turkey; yet many Turks acknowledge that *massacres* were committed against their Armenian population. See also Ch. 2, n. 6.

23 Rafael Lemkin coined the term 'genocide', from the Greek γένος (*genos*, a race or people) and the Latin *caedere* (meaning 'to kill'). Lemkin campaigned successfully to have the concept of genocide accepted as legal reality by the United Nations.

24 Bridgman and Worley (2004), Gewald (1999).

25 Hochschild (1999).

26 For an introduction to the debate on the uniqueness of the Holocaust see Rosenbaum (2000).

27 The examples cited in the text were chosen to illustrate both the range of human complicity in large-scale suffering and the difficulty of attributing causal and moral responsibility in such cases. All involved hundreds of thousands of deaths and left lasting scars on their respective peoples. All are bitterly contested. The atrocious treatment of Chinese soldiers and civilians (and those of other states conquered by Japan from 1937) comprised far more than the Rape

of Nanking, terrible though that was (see Ch. 1, n. 17). Perhaps less well known is the appalling suffering of civilians after General Suharto seized power in Indonesia (1965) and invaded East Timor (1975); events which are estimated to have resulted in at least 600,000 deaths (see Rummel 1994, tables 1.2 and 1.5; see also http://users.erols.com/mwhite28/warstat3.htm#Indonesia). In both cases the locus of moral responsibility seems relatively straightforward: the state forces of Imperial Japan and of General Suharto, respectively. Cruel and deliberate killings, however, do not require bullets. Many have been inflicted by starvation, which may arguably involve more suffering for the victims. Ukrainians experienced immense hardship between the 1920s and 1940s, especially during the great famine of 1932–3 (Vallin et al. 2002), known by some as the Holomodor, or 'hunger murder'. Ireland was devastated by famine following a potato blight in the late 1840s. In both cases human 'perpetrators' (the Stalinist and British authorities, respectively) have been identified, controversially, as playing a decisive role in the disasters. As has been noted elsewhere in this book, however, it is notoriously difficult to determine perpetrator motives (see e.g. English 2007, 161–71, on the Irish famine). Neglect and incompetence, indifference, and determined efforts to worsen the famine can be difficult to disentangle, leaving the field open to ideological bias. As a non-expert in this field, my impression is that the accusation of complicity has considerably more bite in the Ukrainian case than in the Irish one; but neither has been definitively settled.

28 Hardenburg (1912), 28–9.

29 Casement's journal of his visit to the Putumayo is available as Casement (ed. 1997); detailed documentation can be found in Casement (ed. 2003).

30 One of the most notorious atrocities of recent Western history, the Srebrenica massacre took place following the break-up of Yugoslavia. In 1995 Serb forces took over the town of Srebrenica, which was ostensibly under United Nations protection, and killed more than 7,000 Bosnian men and boys. The International Court of Justice has ruled that the event be considered genocide (see http://www.icj-cij.org/docket/files/91/13685.pdf, esp. paras. 278–97). The attack by Islamist terrorists linked to al-Qa'eda on 11 September 2001 ('9/11'), in which passenger jets were flown into iconic American buildings (the World Trade Center and the Pentagon; a third jet crashed in Pennsylvania), killed around 3,000 people.

Bibliography

ABELSON, R. P., DASGUPTA, N., PARK, J., and BANAJI, M. R. (1998), 'Perceptions of the collective other', *Personality and Social Psychology Review*, 2: 243–50.

ADOLPHS, R., TRANEL, D., and DAMASIO, A. R. (2003), 'Dissociable neural systems for recognizing emotions', *Brain and Cognition*, 52: 61–9.

AESCHYLUS (trans. 1922), 'Prometheus Bound'. In *Suppliant Maidens; Persians; Prometheus; Seven against Thebes*, trans. H. W. Smyth. Cambridge, Mass.: Harvard University Press, 209–316.

—— (trans. 1999), *The Oresteia: A New Version by Ted Hughes*. London: Faber & Faber.

AFTANAS, L. I., REVA, N. V., SAVOTINA, L. N., and MAKHNEV, V. P. (2006), 'Neurophysiological correlates of induced discrete emotions in humans: an individually oriented analysis', *Neuroscience and Behavioral Physiology*, 36: 119–30.

AINSLIE, G. (2001), *Breakdown of Will*. New York: Cambridge University Press.

AKUTAGAWA, R. (trans. 2007), *'Rashomon' and Seventeen Other Stories*, trans. J. Rubin. London: Penguin.

ALICKE, M. D. (2000), 'Culpable control and the psychology of blame', *Psychological Bulletin*, 126: 556–74.

AMBROSE, S. H. (1998), 'Late Pleistocene human population bottlenecks, volcanic winter, and differentiation of modern humans', *Journal of Human Evolution*, 34: 623–51.

AMERICAN PSYCHIATRIC ASSOCIATION (2000), *Diagnostic and Statistical Manual of Mental Disorders 4th Edition Text Revision: DSM-IV-TR*. Washington, DC: American Psychiatric Association.

ANDERSON, C. A., DEUSER, W. E., and DENEVE, K. M. (1995), 'Hot temperatures, hostile affect, hostile cognition, and arousal: tests of a general model of affective aggression', *Personality and Social Psychology Bulletin*, 21: 434–8.

APOLLINAIRE, G. (ed. 1964), 'Le Divin Marquis'. In *Les Diables Amoureux*, ed. M. Décaudin. Paris: Gallimard, 178–232.

ARENDT, H. (1963), *Eichmann in Jerusalem: A Report on the Banality of Evil*. London: Faber & Faber.

ARISTOTLE (trans. 1937), 'Parts of Animals'. In *Aristotle: Parts of Animals. Movement of Animals. Progression of Animals*, trans. A. L. Peck and E. S. Forster. Cambridge, Mass.: Harvard University Press, 52–435.

——(trans. 1939), *On the Heavens*, trans. W. K. C. Guthrie. Cambridge, Mass.: Harvard University Press.

ARNDT, J., GREENBERG, J., SCHIMEL, J., PYSZCZYNSKI, T., and SOLOMON, S. (2002), 'To belong or not to belong, that is the question: terror management and identification with gender and ethnicity', *Journal of Personality and Social Psychology*, 83: 26–43.

ARSENEAULT, L., WALSH, E., TRZESNIEWSKI, K., NEWCOMBE, R., CASPI, A., and MOFFITT, T. E. (2006), 'Bullying victimization uniquely contributes to adjustment problems in young children: a nationally representative cohort study', *Pediatrics*, 118: 130–8.

ATRAN, S., AXELROD, R., and DAVIS, R. (2007), 'Sacred barriers to conflict resolution', *Science*, 317: 1039–40.

ATTENBOROUGH, D. and FOTHERGILL, A. (dir.), *Coasts*. In the series *Blue Planet* (2001), shown on BBC 1. Details located at http://www.bbc.co.uk/nature/programmes/tv/blueplanet/.

AUSTIN, P. C., MAMDANI, M. M., JUURLINK, D. N., and HUX, J. E. (2006), 'Testing multiple statistical hypotheses resulted in spurious associations: a study of astrological signs and health', *Journal of Clinical Epidemiology*, 59: 964–9.

AXELROD, R. (1984), *Evolution of Cooperation*. New York: Basic Books.

——(2003), 'The evolution of ethnocentric behavior'. Presented at the Midwest Political Science Convention, Chicago, IL. Located at http://wwwpersonal.umich.edu/~axe/research/AxHamm_Ethno.pdf.

AZIZ, Q., THOMPSON, D. G., NG, V. W., HAMDY, S., SARKAR, S., BRAMMER, M. J., BULLMORE, E. T., HOBSON, A., TRACEY, I., GREGORY, L., SIMMONS, A., and WILLIAMS, S. C. (2000), 'Cortical processing of human somatic and visceral sensation', *Journal of Neuroscience*, 20: 2657–63.

BALAKIAN, P. (1998), *Black Dog of Fate: A Memoir*. New York: Broadway.

BALLIF-SPANVILL, B., CLAYTON, C. J., and HENDRIX, S. B. (2003), 'Gender, types of conflict, and individual differences in the use of violent and peaceful strategies among children who have and have not witnessed interparental violence', *American Journal of Orthopsychiatry*, 73: 141–53.

BANDURA, A. (2002), 'Selective moral disengagement in the exercise of moral agency', *Journal of Moral Education*, 31: 101–19.

BAR, M. (2007), 'The proactive brain: using analogies and associations to generate predictions', *Trends in Cognitive Sciences*, 11: 280–9.

BARAK, G. (2003), *Violence and Nonviolence: Pathways to Understanding*. London: Sage.

BARRETT, L. F., LINDQUIST, K. A., and GENDRON, M. (2007), 'Language as context for the perception of emotion', *Trends in Cognitive Sciences*, 11: 327–32.

BARRETT, L. F., MESQUITA, B., OCHSNER, K. N., and GROSS, J. J. (2007), 'The experience of emotion', *Annual Review of Psychology*, 58: 373–403.

BARROZO, P. (2008), 'Punishing cruelly: punishment, cruelty, and mercy', *Criminal Law and Philosophy*, 2: 67–84.

BARRY, B. (2001), *Culture and Equality: An Egalitarian Critique of Multiculturalism*. Cambridge: Polity.

BAUMEISTER, R. F. (2001), *Evil: Inside Human Violence and Cruelty*. New York: Owl Books.

BECHARA, A., DAMASIO, H., and DAMASIO, A. R. (2000), 'Emotion, decision making and the orbitofrontal cortex', *Cerebral Cortex*, 10: 295–307.

—— —— TRANEL, D., and DAMASIO, A. R. (2005), 'The Iowa Gambling Task and the somatic marker hypothesis: some questions and answers', *Trends in Cognitive Sciences*, 9: 159–62; discussion, pp. 162–4.

BECK, A. T. (1999), *Prisoners of Hate: The Cognitive Basis of Anger, Hostility, and Violence*. New York: HarperCollins.

BENARROCH, E. E. (1997), *The Central Autonomic Network: Functional Organization and Clinical Correlations*. Armonk, NY: Futura.

BENDER, S., HELLWIG, S., RESCH, F., and WEISBROD, M. (2007), 'Am I safe? The ventrolateral prefrontal cortex "detects" when an unpleasant event does not occur', *NeuroImage*, 38: 367–85.

BENTHAM, J. (ed. 2007), *An Introduction to the Principles of Morals and Legislation*, 1823 edn. Mineola, NY: Dover.

BERG, K. (trans. 1938), *The Sadist*, trans. O. Illner and G. Godwin. London: Acorn Press.

BERKOWITZ, L. (1999), 'Evil is more than banal: situationism and the concept of evil', *Personality and Social Psychology Review*, 3: 246–53.

BERNS, G. S., LAIBSON, D., and LOEWENSTEIN, G. (2007), 'Intertemporal choice—toward an integrative framework', *Trends in Cognitive Sciences*, 11: 482–8.

BERNSTEIN, M. J., YOUNG, S. G., and HUGENBERG, K. (2007), 'The cross-category effect: mere social categorization is sufficient to elicit an own-group bias in face recognition', *Psychological Science*, 18: 706–12.

BIRCH, S. A. J. and BLOOM, P. (2007), 'The curse of knowledge in reasoning about false beliefs', *Psychological Science*, 18: 382–6.

BLACK, E. (2004), *War Against the Weak: Eugenics and America's Campaign to Create a Master Race*. New York: Thunder's Mouth.

BLACKMORE, S. (2000), *The Meme Machine*. New York: Oxford University Press.

BLAIR, A., *Full text: Tony Blair speech on terror*. Located at http://news.bbc.co.uk/1/hi/uk/4689363.stm.

BLAIR, R. J. (2005), 'Responding to the emotions of others: dissociating forms of empathy through the study of typical and psychiatric populations', *Consciousness and Cognition*, 14: 698–718.

BLAIR, R. J. R., JONES, L., CLARK, F., and SMITH, M. (1997), 'The psychopathic individual: a lack of responsiveness to distress cues?' *Psychophysiology*, 34: 192–8.

BLOCK, N. (1995), 'On a confusion about a function of consciousness', *Behavioral and Brain Sciences*, 18: 227–87.

BLOOM, M. (2005), *Dying to Kill: the Allure of Suicide Terror*. New York: Columbia University Press.

BODEN-ALBALA, B., LITWAK, E., ELKIND, M. S., RUNDEK, T., and SACCO, R. L. (2005), 'Social isolation and outcomes post stroke', *Neurology*, 64: 1888–92.

BOITEN, F. (1996), 'Autonomic response patterns during voluntary facial action', *Psychophysiology*, 33: 123–31.

BOKLAGE, C. E. (1990), 'Survival probability of human conceptions from fertilization to term', *International Journal of Fertility (Stockholm)*, 35: 75–94.

BORG, J. S., HYNES, C., VAN HORN, J., GRAFTON, S., and SINNOTT-ARMSTRONG, W. (2006), 'Consequences, action, and intention as factors in moral judgments: an fMRI investigation', *Journal of Cognitive Neuroscience*, 18: 803–17.

BOROWITZ, A. (2005), *Terrorism for Self-glorification: The Herostratos Syndrome*. Kent, Ohio: Kent State University Press.

BOTVINICK, M. M., COHEN, J. D., and CARTER, C. S. (2004), 'Conflict monitoring and anterior cingulate cortex: an update', *Trends in Cognitive Sciences*, 8: 539–46.

BOURKE, J. (2000), *An Intimate History of Killing: Face-to-face Killing in Twentieth-Century Warfare*. London: Granta.

BRAVER, T. S., BARCH, D. M., GRAY, J. R., MOLFESE, D. L., and SNYDER, A. (2001), 'Anterior cingulate cortex and response conflict: effects of frequency, inhibition and errors', *Cerebral Cortex*, 11: 825–36.

BRECHT, M. and SCHMITZ, D. (2008), 'Rules of plasticity', *Science*, 319: 39–40.

BREWER, M. B. (2007), 'The importance of being We: human nature and intragroup relations', *American Psychologist*, 62: 728–38.

BRIDGMAN, J. and WORLEY, L. J. (2004), 'Genocide of the Hereros'. In S. Totten, W. S. Parsons, and I. W. Charny (eds.), *Century of Genocide: Critical Essays and Eyewitness Accounts*, 2nd edn. New York: Routledge, 15–51.

BROGDEN, M. (2001), *Geronticide: Killing the Elderly*. London: Jessica Kingsley.

BROUSSARD, D. L. and ALTSCHULER, S. M. (2000), 'Brainstem viscerotopic organization of afferents and efferents involved in the control of swallowing', *American Journal of Medicine*, 108: S79–86.

BROWNE, A. and FINKELHOR, D. (1986), 'Impact of child sexual abuse: a review of the research', *Psychological Bulletin*, 99: 66–77.

BROWNING, C. (1991), *Ordinary Men: Reserve Police Battalion 101 and the Final Solution in Poland*. New York: HarperCollins.

BUFKIN, J. L. and LUTTRELL, V. R. (2005), 'Neuroimaging studies of aggressive and violent behavior: current findings and implications for criminology and criminal justice', *Trauma, Violence, and Abuse*, 6: 176–91.

BULLER, K. M. (2003), 'Neuroimmune stress responses: reciprocal connections between the hypothalamus and the brainstem', *Stress*, 6: 11–17.

BURKE, J. (2004), *Al Qaeda: The True Story of Radical Islam*. London: I. B. Tauris.

BURNS, R. (ed. 1993), 'To a louse: on seeing one on a lady's bonnet at church'. In *Robert Burns: Selected Poems*, ed. C. McGuirk. London: Penguin, 85–6.

BURRIS, C. T. and REMPEL, J. K. (2004), ' "It's the end of the world as we know it": threat and the spatial-symbolic self', *Journal of Personality and Social Psychology*, 86: 19–42.

BUSHMAN, B. J., RIDGE, R. D., DAS, E., KEY, C. W., and BUSATH, G. L. (2007), 'When God sanctions killing: effect of scriptural violence on aggression', *Psychological Science*, 18: 204–7.

BUSS, D. M. (2005), *The Murderer Next Door: Why the Mind Is Designed to Kill*. New York: Penguin.

—— and DUNTLEY, J. D. (2005), 'The plausibility of adaptations for homicide'. In P. Carruthers, S. Laurence, and S. Stich (eds.), *The Innate Mind: Structure and Contents*. Oxford: Oxford University Press, 291–304.

BUTLER, T., PAN, H., TUESCHER, O., ENGELIEN, A., GOLDSTEIN, M., EPSTEIN, J., WEISHOLTZ, D., ROOT, J. C., PROTOPOPESCU, X., CUNNINGHAM-BUSSEL, A. C., CHANG, L., XIE, X. H., CHEN, Q., PHELPS, E. A., LEDOUX, J. E., STERN, E., and SILBERSWEIG, D. A. (2007), 'Human fear-related motor neurocircuitry', *Neuroscience*, 150: 1–7.

BUZSAKI, G. (2006), *Rhythms of the Brain*. New York: Oxford University Press.

BUZSAKI, G., KAILA, K., and RAICHLE, M. (2007), 'Inhibition and brain work', *Neuron*, 56: 771–83.

BYRNE, J. H. and ROBERTS, J. L., eds. (2004), *From Molecules to Networks: An Introduction to Cellular and Molecular Neuroscience*. London: Academic Press.

CACIOPPO, J. T. and HAWKLEY, L. C. (2003), 'Social isolation and health, with an emphasis on underlying mechanisms', *Perspectives in Biology and Medicine*, 46: S39–52.

CAMERON, O. G. (2001), 'Interoception: the inside story—a model for psychosomatic processes', *Psychosomatic Medicine*, 63: 697–710.

CAMPBELL, H. L., TIVARUS, M. E., HILLIER, A., and BEVERSDORF, D. Q. (2008), 'Increased task difficulty results in greater impact of noradrenergic modulation of cognitive flexibility', *Pharmacology Biochemistry and Behavior*, 88: 222–9.

CARD, C. (2002), *The Atrocity Paradigm: A Theory of Evil*. Oxford: Oxford University Press.

CASE, T. I., REPACHOLI, B. M., and STEVENSON, R. J. (2006), 'My baby doesn't smell as bad as yours: the plasticity of disgust', *Evolution and Human Behavior*, 27: 357–65.

CASEMENT, R. (ed. 1997), *The Amazon Journal of Roger Casement*, edited and with an Introduction by Angus Mitchell. London: Anaconda.

—— (ed. 2003), *Sir Roger Casement's Heart of Darkness: The 1911 Documents; Introduction, Commentary and Footnotes by Angus Mitchell*. Dublin: Irish Manuscripts Commission.

CASTANO, E. and GINER-SOROLLA, R. (2006), 'Not quite human: infrahumanization in response to collective responsibility for intergroup killing', *Journal of Personality and Social Psychology*, 90: 804–18.

CASTELL, D. O., WOOD, J. D., FRIELING, T., WRIGHT, F. S., and VIETH, R. F. (1990), 'Cerebral electrical potentials evoked by balloon distention of the human esophagus', *Gastroenterology*, 98: 662–6.

CAVAFY, C. P. (trans. 2007), *The Collected Poems*, trans. Evangelos Sachperoglou. Oxford: Oxford University Press.

CHAMPLIN, E. (2003), *Nero*. Cambridge, Mass.: Belknap.

CHANG, I. (1998), *The Rape of Nanking: The Forgotten Holocaust of World War II*. London: Penguin.

CHEN, C. Y., MUGGLETON, N. G., JUAN, C. H., TZENG, O. J. L., and HUNG, D. L. (2008), 'Time pressure leads to inhibitory control deficits in impulsive violent offenders', *Behavioural Brain Research*, 187: 483–8.

CHIROT, D. and McCAULEY, C. (2006), *Why Not Kill Them All? The Logic and Prevention of Mass Political Murder*. Princeton: Princeton University Press.

CHOMSKY, N. (1957), *Syntactic Structures*. The Hague: Mouton.

CHURCHLAND, P. S. (1989), *Neurophilosophy: Toward a Unified Science of the Mind/Brain*. Cambridge, Mass.: MIT Press.

CLORE, G. L. and HUNTSINGER, J. R. (2007), 'How emotions inform judgement and regulate thought', *Trends in Cognitive Sciences*, 11: 393–9.

COATES, A. J. (1997), *The Ethics of War*. Manchester: Manchester University Press.

COHEN, G. L., SHERMAN, D. K., BASTARDI, A., HSU, L., McGOEY, M., and ROSS, L. (2007), 'Bridging the partisan divide: self-affirmation reduces ideological closed-mindedness and inflexibility in negotiation', *Journal of Personality and Social Psychology*, 93: 415–30.

COLLET, C., VERNET-MAURY, E., DELHOMME, G., and DITTMAR, A. (1997), 'Autonomic nervous system response patterns specificity to basic emotions', *Journal of the Autonomic Nervous System*, 62: 45–57.

COMISION PARA EL ESCLARACIMIENTO HISTORICO (last accessed 29 May 2008), *Guatemala: Memoria del Silencio* [Memory of Silence]. Report of the Comision para el Esclaracimiento Historico [CEH; Commission for Historical Clarification]. Located at http://shr.aaas.org/guatemala/ceh/report/english/toc.html.

CONROY, J. (2001), *Unspeakable Acts, Ordinary People: The Dynamics of Torture*. London: Vision.

COOPER, J. (2007), *Cognitive Dissonance: Fifty Years of a Classic Theory*. London: Sage.

CORNWELL, J. (2003), *Darwin's Angel: An Angelic Riposte to 'The God Delusion'*. London: Viking.

CORTES, B. P., DEMOULIN, S., RODRIGUEZ, R. T., RODRIGUEZ, A. P., and LEYENS, J. P. (2005), 'Infrahumanization or familiarity? Attribution of uniquely human emotions to the self, the ingroup, and the outgroup', *Personality and Social Psychology Bulletin*, 31: 243–53.

COUPPIS, M. H. and KENNEDY, C. H. (2008), 'The rewarding effect of aggression is reduced by nucleus accumbens dopamine receptor antagonism in mice', *Psychopharmacology*, 197: 449–56.

COX, J. J., REIMANN, F., NICHOLAS, A. K., THORNTON, G., ROBERTS, E., SPRINGELL, K., KARBANI, G., JAFRI, H., MANNAN, J., RAASHID, Y., AL-GAZALI, L., HAMAMY, H., VALENTE, E. M., GORMAN, S., WILLIAMS, R., MCHALE, D. P., WOOD, J. N., GRIBBLE, F. M., and WOODS, C. G. (2006), 'An SCN9A channelopathy causes congenital inability to experience pain', *Nature*, 444: 894–8.

CROSS, P. A. and MATHESON, K. (2006), 'Understanding sadomasochism: an empirical examination of four perspectives', *Journal of Homosexuality*, 50: 133–66.

CROYLE, R. T. and COOPER, J. (1983), 'Dissonance arousal: physiological evidence', *Journal of Personality and Social Psychology*, 45: 782–91.

CSIKSZENTMIHALYI, M. (2002), *Flow: The Classic Work on How to Achieve Happiness*. London: Rider.

CUNLIFFE, B. (2006), 'The Roots of Warfare', in M. Jones and A. Fabian (eds.), *Conflict*. Cambridge: Cambridge University Press, 63–81.

CUNNINGHAM, W. A. and ZELAZO, P. D. (2007), 'Attitudes and evaluations: a social cognitive neuroscience perspective', *Trends in Cognitive Sciences*, 11: 97–104.

CURTIS, A. (dir.), *The Power of Nightmares*. Shown on BBC 2 (2004). Details located at http://news.bbc.co.uk/1/hi/programmes/4202741.stm.

Curtis, V., Aunger, R., and Rabie, T. (2004), 'Evidence that disgust evolved to protect from risk of disease', *Philosophical Transactions of the Royal Society Series B: Biological Sciences*, 271: S131–3.

Dadds, M. R., Whiting, C., and Hawes, D. J. (2006), 'Associations among cruelty to animals, family conflict, and psychopathic traits in childhood', *Journal of Interpersonal Violence*, 21: 411–29.

Dadrian, V. N. (1995), *The History of the Armenian Genocide: Ethnic Conflict from the Balkans to Anatolia to the Caucasus*. Oxford: Berghahn.

Dallaire, R. (2004), *Shake Hands with the Devil: The Failure of Humanity in Rwanda*. London: Arrow.

Daly, M. and Wilson, M. (1998), *The Truth About Cinderella*. New Haven: Yale University Press.

Damasio, A. (1996), *Descartes' Error: Emotion, Reason and the Human Brain*. London: Papermac.

——(2000), *The Feeling of What Happens: Body, Emotion and the Making of Consciousness*. London: Heinemann.

——(2003), *Looking for Spinoza: Joy, Sorrow and the Feeling Brain*. London: Heinemann.

Damasio, A. R., Grabowski, T. J., Bechara, A., Damasio, H., Ponto, L. L. B., Parvizi, J., and Hichwa, R. D. (2000), 'Subcortical and cortical brain activity during the feeling of self-generated emotions', *Nature Neuroscience*, 3: 1049–56.

Danner, M. (2005), *The Massacre at El Mozote: A Parable of the Cold War*. London: Granta.

Danziger, N., Prkachin, K. M., and Willer, J. C. (2006), 'Is pain the price of empathy? The perception of others' pain in patients with congenital insensitivity to pain', *Brain*, 129: 2494–507.

Darby, B. W. and Jeffers, D. (1988), 'The effects of defendant and juror attractiveness on simulated courtroom trial decisions', *Social Behavior and Personality*, 16: 39–50.

Darley, J. M. and Batson, C. D. (1973), ' "From Jerusalem to Jericho": a study of situational and dispositional variables in helping behaviour', *Journal of Personality and Social Psychology*, 27: 100–8.

Darwin, C. ed. (1999), *The Expression of the Emotions in Man and Animals*, ed. P. Ekman. London: HarperCollins.

Davey, G. C. (1994), 'Self-reported fears to common indigenous animals in an adult UK population: the role of disgust sensitivity', *British Journal of Psychology*, 85: 541–54.

—— Bickerstaffe, S., and Macdonald, B. A. (2006), 'Experienced disgust causes a negative interpretation bias: a causal role for disgust in anxious psychopathology', *Behaviour Research and Therapy*, 44: 1375–84.

DAVEY, G. C., CAVANAGH, K., and LAMB, A. (2003), 'Differential aversive outcome expectancies for high- and low-predation fear-relevant animals', *Journal of Behavior Therapy and Experimental Psychiatry*, 34: 117–28.

——— FORSTER, L., and MAYHEW, G. (1993), 'Familial resemblances in disgust sensitivity and animal phobias', *Behaviour Research and Therapy*, 31: 41–50.

——— McDONALD, A. S., HIRISAVE, U., PRABHU, G. G., IWAWAKI, S., JIM, C. I., MERCKELBACH, H., DE JONG, P. J., LEUNG, P. W., and REIMANN, B. C. (1998), 'A cross-cultural study of animal fears', *Behaviour Research and Therapy*, 36: 735–50.

DAWKINS, R. (1989), *The Selfish Gene*, new edn. Oxford: Oxford University Press.

——— (2006), *The God Delusion*. London: Bantam Press.

DAY, L. H. (1984), 'Death from non-war violence: an international comparison', *Social Science and Medicine*, 19: 917–27.

DEACON, T. W. (1997), *The Symbolic Species: The Co-evolution of Language and the Human Brain*. London: Allen Lane.

DEBOER, S. L. and MADDOW, C. L. (2002), 'Emergency care of the crucifixion victim', *Accident and Emergency Nursing*, 10: 235–9.

DECLERCK, C. H., BOONE, C., and DE BRABANDER, B. (2006), 'On feeling in control: a biological theory for individual differences in control perception', *Brain and Cognition*, 62: 143–76.

DELEUZE, G. (trans. 1971), *Sacher-Masoch: An Interpretation*, trans. J. McNeil. London: Faber & Faber.

DENNETT, D. C. (1989), *The Intentional Stance*, new edn. Cambridge, Mass.: MIT Press.

——— (1995), *Darwin's Dangerous Idea*. New York: Simon & Schuster.

——— (2003a), *Consciousness Explained*. London: Penguin.

——— (2003b), *Freedom Evolves*. London: Allen Lane.

——— (2006), *Breaking the Spell: Religion as a Natural Phenomenon*. London: Allen Lane.

DESCARTES, R. (trans. 1988), *Selected Philosophical Writings*, trans. J. Cottingham, R. Stoothoff, and D. Murdoch. Cambridge: Cambridge University Press.

DESIMONE, R. (1996), 'Neural mechanisms for visual memory and their role in attention', *Proceedings of the National Academy of Sciences of the United States of America*, 93: 13494–9.

DESTENO, D., DASGUPTA, N., BARTLETT, M. Y., and CAJDRIC, A. (2004), 'Prejudice from thin air', *Psychological Science*, 15: 319–24.

DEVINE, D. J., CLAYTON, L. D., DUNFORD, B. B., SEYING, R., and PRYCE, J. (2001), 'Jury decision making: 45 years of empirical research on deliberating groups', *Psychology, Public Policy, and Law*, 7: 622–727.

DICKERSON, S. S. and KEMENY, M. E. (2004), 'Acute stressors and cortisol responses: a theoretical integration and synthesis of laboratory research', *Psychological Bulletin*, 130: 355–91.

DILILLO, D. and DAMASHEK, A. (2003), 'Parenting characteristics of women reporting a history of childhood sexual abuse', *Child Maltreatment*, 8: 319–33.

DOUGLAS, E. M. (2006), 'Familial violence socialization in childhood and later life approval of corporal punishment: a cross-cultural perspective', *American Journal of Orthopsychiatry*, 76: 23–30.

DOUGLAS, M. (2002), *Purity and Danger*. London: Routledge.

DREBER, A., RAND, D. G., FUDENBERG, D., and NOWAK, M. A. (2008), 'Winners don't punish', *Nature*, 452: 348–51.

DUGATKIN, L. A. (2006), *The Altruism Equation: Seven Scientists Search for the Origins of Goodness*. Princeton: Princeton University Press.

DUMAS, R. and TESTÉ, B. (2006), 'The influence of criminal facial stereotypes on juridic judgments', *Swiss Journal of Psychology*, 65: 237–44.

DUNBAR, R. (1997), *Grooming, Gossip and the Evolution of Language*. London: Faber & Faber.

DUNN, B. D., DALGLEISH, T., and LAWRENCE, A. D. (2006), 'The somatic marker hypothesis: a critical evaluation', *Neuroscence and Biobehavioral Reviews*, 30: 239–71.

DUPOUX, E. and JACOB, P. (2007), 'Universal moral grammar: a critical appraisal', *Trends in Cognitive Sciences*, 11: 373–8.

DUTTON, D. G. (2007), *The Psychology of Genocide, Massacres, and Extreme Violence*. Westport, Conn.: Praeger Security International.

EDELMAN, G. M. (1987), *Neural Darwinism: The Theory of Neuronal Group Selection*. New York: Basic Books.

EDWARDS, C. M. (1988), 'Chemotherapy induced emesis—mechanisms and treatment: a review', *Journal of the Royal Society of Medicine*, 81: 658–62.

EGAN, L. C., SANTOS, L. R., and BLOOM, P. (2007), 'The origins of cognitive dissonance: evidence from children and monkeys', *Psychological Science*, 18: 978–83.

EGNER, T. and HIRSCH, J. (2005), 'Cognitive control mechanisms resolve conflict through cortical amplification of task-relevant information', *Nature Neuroscience*, 8: 1784–90.

—— ETKIN, A., GALE, S., and HIRSCH, J. (2008), 'Dissociable neural systems resolve conflict from emotional versus nonemotional distracters', *Cerebral Cortex*, 18: 1475–84.

EICKHOFF, S. B., LOTZE, M., WIETEK, B., AMUNTS, K., ENCK, P., and ZILLES, K. (2006), 'Segregation of visceral and somatosensory afferents: an fMRI and cytoarchitectonic mapping study', *NeuroImage*, 31: 1004–14.

EKMAN, P., SORENSON, E. R., and FRIESEN, W. V. (1969), 'Pan-cultural elements in facial displays of emotion', *Science*, 164: 86–8.

ELLIOTT, A. J. and DEVINE, P. G. (1994), 'On the motivational nature of cognitive dissonance: dissonance as psychological discomfort', *Journal of Personality and Social Psychology*, 67: 382–94.

ELLIOTT, C. (1996), *The Rules of Insanity: Moral Responsibility and the Mentally Ill Offender*. Albany, NY: State University of New York Press.

ENGLISH, R. (2007), *Irish Freedom: The History of Nationalism in Ireland*. London: Pan.

EPLEY, N., WAYTZ, A., and CACIOPPO, J. T. (2007), 'On seeing human: a three-factor theory of anthropomorphism', *Psychological Review*, 114: 864–86.

ERTEM, I. O., LEVENTHAL, J. M., and DOBBS, S. (2000), 'Intergenerational continuity of child physical abuse: how good is the evidence?' *Lancet*, 356: 814–19.

ESSLEN, M., PASCUAL-MARQUI, R. D., HELL, D., KOCHI, K., and LEHMANN, D. (2004), 'Brain areas and time course of emotional processing', *NeuroImage*, 21: 1189–1203.

EURIPIDES (trans. 1963), 'Medea'. In *Medea and Other Plays*, trans. P. Vellacott. Harmondsworth: Penguin, 17–62.

—— (trans. 1973), 'The Bacchae'. In *The Bacchae and Other Plays*, trans. P. Vellacott. Harmondsworth: Penguin, 191–244.

FAHY, F. L., RICHES, I. P., and BROWN, M. W. (1993), 'Neuronal activity related to visual recognition memory: long-term memory and the encoding of recency and familiarity information in the primate anterior and medial inferior temporal and rhinal cortex', *Experimental Brain Research*, 96: 457–72.

FASCHING, D. (1992), *Narrative Theology after Auschwitz: From Alienation to Ethics*. Minneapolis: Fortress Press.

FAULKNER, J., SCHALLER, M., PARK, J. H., and DUNCAN, L. A. (2004), 'Evolved disease-avoidance mechanisms and contemporary xenophobic attitudes', *Group Processes and Intergroup Relations*, 7: 333–53.

FEHR, E. and GACHTER, S. (2002), 'Altruistic punishment in humans', *Nature*, 415: 137–40.

—— and ROCKENBACH, B. (2004), 'Human altruism: economic, neural, and evolutionary perspectives', *Current Opinion in Neurobiology*, 14: 784–90.

FEIN, H. (1990), 'Genocide: a sociological perspective', *Current Sociology*, 38: 1 (special issue).

FELLEMAN, D. J. and VAN ESSEN, D. C. (1991), 'Distributed hierarchical processing in the primate cerebral cortex', *Cerebral Cortex*, 1: 1–47.

FERGUSON, M. J., BARGH, J. A., and NAYAK, D. A. (2005), 'After-affects: how automatic evaluations influence the interpretation of subsequent, unrelated stimuli', *Journal of Experimental Social Psychology*, 41: 182–91.

FESSLER, D. M. T. (2002), 'Reproductive immunosuppression and diet: an evolutionary perspective on pregnancy sickness and meat consumption', *Current Anthropology*, 43: 19–39.

——— and HALEY, K. J. (2006), 'Guarding the perimeter: the outside–inside dichotomy in disgust and bodily experience', *Cognition and Emotion*, 20: 3–19.

——— ENG, S. J., and NAVARRETE, C. D. (2005), 'Elevated disgust sensitivity in the first trimester of pregnancy: evidence supporting the compensatory prophylaxis hypothesis', *Evolution and Human Behavior*, 26: 344–51.

FESTINGER, L. (1957), *A Theory of Cognitive Dissonance.* New York: Row, Peterson & Co.

FINE, C. (2007), *A Mind of Its Own: How Your Brain Distorts and Deceives.* Cambridge: Icon.

FINKELHOR, D. (1990), 'Early and long-term effects of child sexual abuse: an update', *Professional Psychology: Research and Practice*, 21: 325–30.

FISCHER, A. H. and ROSEMAN, I. J. (2007), 'Beat them or ban them: the characteristics and social functions of anger and contempt', *Journal of Personality and Social Psychology*, 93: 103–15.

FISHBACH, A. and TROPE, Y. (2005), 'The substitutability of external control and self-control', *Journal of Experimental Social Psychology*, 41: 256–70.

FISK, R., 'What do you say to a man whose family is buried under the rubble?' *The Independent*, 9 Aug. 2006.

FLORIAN, V., MIKULINCER, M., and HIRSCHBERGER, G. (2002), 'The anxiety-buffering function of close relationships: evidence that relationship commitment acts as a terror management mechanism', *Journal of Personality and Social Psychology*, 82: 527–42.

FOGASSI, L., FERRARI, P. F., GESIERICH, B., ROZZI, S., CHERSI, F., and RIZZOLATTI, G. (2005), 'Parietal lobe: from action organization to intention understanding', *Science*, 308: 662–7.

FOOT, P. (1978), 'The problem of abortion and the doctrine of the double effect'. In *Virtues and Vices and Other Essays in Moral Philosophy.* Oxford: Basil Blackwell, 19–32.

FORGAS, P. (1998), 'On feeling good and getting your way: mood effects on negotiator cognition and bargaining strategies', *Journal of Personality and Social Psychology*, 74: 565–77.

FOSTER, D. P. and YOUNG, H. P. (2001), 'On the impossibility of predicting the behavior of rational agents', *Proceedings of the National Academy of Sciences of the United States of America*, 98: 12848–53.

FOXE, J. (ed. *c*.1910), *Book of Martyrs*, ed. C. H. H. Wright; n.d. London: J. A. Kensit.

FRANKFURTER, D. (2006), *Evil Incarnate: Rumors of Demonic Conspiracy and Satanic Abuse in History*. Princeton: Princeton University Press.

FRICK, P. J. and DICKENS, C. (2006), 'Current perspectives on conduct disorder', *Current Psychiatry Reports*, 8: 59–72.

FRISTON, K. (2005), 'A theory of cortical responses', *Philosophical Transactions of the Royal Society Series B: Biological Sciences*, 360: 815–36.

FRITH, C. B., FRITH, D. W., and BARNES, E. (2004), *The Bowerbirds*. Oxford: Oxford University Press.

FRITH, C. D. (2007), *Making up the Mind*. Oxford: Blackwell.

—— and FRITH, U. (2006), 'The neural basis of mentalizing', *Neuron*, 50: 531–4.

FROMM, E. (1975), *The Anatomy of Human Destructiveness*. Greenwich, Conn.: Fawcett.

GAILLIOT, M. T., BAUMEISTER, R. F., DEWALL, C. N., MANER, J. K., PLANT, E. A., TICE, D. M., BREWER, L. E., and SCHMEICHEL, B. J. (2007), 'Self-control relies on glucose as a limited energy source: willpower is more than a metaphor', *Journal of Personality and Social Psychology*, 92: 325–36.

GALLESE, V., FADIGA, L., FOGASSI, L., and RIZZOLATTI, G. (1996), 'Action recognition in the premotor cortex', *Brain*, 119: 593–609.

GAMBETTA, D., ed. (2005), *Making Sense of Suicide Missions*. New York: Oxford University Press.

GAUTHIER, I., SKUDLARSKI, P., GORE, J. C., and ANDERSON, A. W. (2000), 'Expertise for cars and birds recruits brain areas involved in face recognition', *Nature Neuroscience*, 3: 191–7.

GEERTZ, C. (1973), *The Interpretation of Cultures: Selected Essays*. New York: Basic Books.

GERGELY, G., NADASDY, Z., CSIBRA, G., and BIRO, S. (1995), 'Taking the intentional stance at 12 months of age', *Cognition*, 56: 165–93.

GERSHOFF, E. T. (2002), 'Corporal punishment by parents and associated child behaviors and experiences: a meta-analytic and theoretical review', *Psychological Bulletin*, 128: 539–79.

GESCH, C. B., HAMMOND, S. M., HAMPSON, S. E., EVES, A., and CROWDER, M. J. (2002), 'Influence of supplementary vitamins, minerals and essential fatty acids on the antisocial behaviour of young adult prisoners. Randomised, placebo-controlled trial.' *British Journal of Psychiatry*, 181: 22–8.

GEWALD, J. B. (1999), *Herero Heroes: A Socio-political History of the Herero of Namibia, 1890–1923*. Oxford: James Currey.

GHERMAN, A , CHEN, P. E., TESLOVICH, T. M., STANKIEWICZ, P., WITHERS, M., KASHUK, C. S., CHAKRAVARTI, A., LUPSKI, J. R., CUTLER, D. J., and KATSANIS, N. (2007), 'Population bottlenecks as a potential major shaping force of human genome architecture', *PLoS Genetics*, 3, p. e119. DOI: http://dx.doi.org/10.1371/journal.pgen.0030119.

GINGES, J., ATRAN, S., MEDIN, D., and SHIKAKI, K. (2007), 'Sacred bounds on rational resolution of violent political conflict', *Proceedings of the National Academy of Sciences of the United States of America*, 104: 7357–60.

GIRARD, R. (trans. 2005), *Violence and the Sacred*, trans. P. Gregory. London: Continuum.

GLOVER, J. (2001), *Humanity: A Moral History of the Twentieth Century*. London: Pimlico.

GOLDBERG, Y. P., MACFARLANE, J., MACDONALD, M. L., THOMPSON, J., DUBE, M. P., MATTICE, M., FRASER, R., YOUNG, C., HOSSAIN, S., PAPE, T., PAYNE, B., RADOMSKI, C., DONALDSON, G., IVES, E., COX, J., YOUNGHUSBAND, H. B., GREEN, R., DUFF, A., BOLTSHAUSER, E., GRINSPAN, G. A., DIMON, J. H., SIBLEY, B. G., ANDRIA, G., TOSCANO, E., KERDRAON, J., BOWSHER, D., PIMSTONE, S. N., SAMUELS, M. E., SHERRINGTON, R., and HAYDEN, M. R. (2007), 'Loss-of-function mutations in the Nav1.7 gene underlie congenital indifference to pain in multiple human populations', *Clinical Genetics*, 71: 311–19.

GOLDHAGEN, D. J. (1997), *Hitler's Willing Executioners: Ordinary Germans and the Holocaust*. London: Abacus.

GOUREVITCH, P. (2000), *We Wish to Inform You That Tomorrow We Will Be Killed with Our Families*. London: Picador.

DE GRAAF, J., WANN, D., and NAYLOR, T. H. (2001), *Affluenza: The All-consuming Epidemic*. San Francisco: Berrett-Koehler.

GRAY, N. S., WATT, A., HASSAN, S., and MACCULLOCH, M. J. (2003), 'Behavioral indicators of sadistic sexual murder predict the presence of sadistic sexual fantasy in a normative sample', *Journal of Interpersonal Violence*, 18: 1018–34.

GREEN, T. (2007), *Inquisition: The Reign of Fear*. London: Macmillan.

GREENBERG, J., SOLOMON, S., and PYSZCZYNSKI, T. (1997), 'Terror management theory of self-esteem and cultural worldviews: empirical assessments and conceptual refinements'. In M. P. Zanna (ed.), *Advances in Experimental Social Psychology: Volume 29*. New York: Academic Press, 61–139.

GREENE, J. D., NYSTROM, L. E., ENGELL, A. D., DARLEY, J. M., and COHEN, J. D. (2004), 'The neural bases of cognitive conflict and control in moral judgment', *Neuron*, 44: 389–400.

GREENE, J. D., SOMMERVILLE, R.B., NYSTROM, L.E., DARLEY, J.M., and COHEN, J.D. (2001), 'An fMRI investigation of emotional engagement in moral judgment', *Science*, 293: 2105–8.

GREGORY, L. J., YAGUEZ, L., WILLIAMS, S. C., ALTMANN, C., COEN, S. J., NG, V., BRAMMER, M. J., THOMPSON, D. G., and AZIZ, Q. (2003), 'Cognitive modulation of the cerebral processing of human oesophageal sensation using functional magnetic resonance imaging', *Gut*, 52: 1671–7.

GROSS, J. T. (2003), *Neighbors: The Destruction of the Jewish Community in Jedwabne, Poland*. Princeton: Princeton University Press.

GU, X. and HAN, S. (2007), 'Attention and reality constraints on the neural processes of empathy for pain', *NeuroImage*, 36: 256–67.

HACKER, P. M. S. and BENNETT, M. R. (2003), *Philosophical Foundations of Neuroscience*. Oxford: Blackwell.

HAIDT, J. (2007), 'The new synthesis in moral psychology', *Science*, 316: 998–1002.

—— McCAULEY, C., and ROZIN, P. (1994), 'Individual differences in sensitivity to disgust: a scale sampling seven domains of disgust elicitors', *Personality and Individual Differences*, 16: 701–13.

—— ROZIN, P., McCAULEY, C., and IMADA, S. (1997), 'Body, psyche and culture: the relationship between disgust and morality', *Psychology and Developing Societies*, 9: 107–31.

HAJCAK, G. and FOTI, D. (2008), 'Errors are aversive: defensive motivation and the error-related negativity', *Psychological Science*, 19: 103–8.

HAMILTON, W. D. (1964), 'The genetical evolution of social behavior, I and II', *Theoretical Biology*, 7: 1–52.

HARDENBURG, W. V. (1912), *The Putumayo: The Devil's Paradise*. London: T. Fisher Unwin.

HARE, R. D. (1999), *Without Conscience: The Disturbing World of the Psychopaths among Us*. London: Guilford Press.

HARRIS, L. T. and FISKE, S. T. (2006), 'Dehumanizing the lowest of the low: neuroimaging responses to extreme out-groups', *Psychological Science*, 17: 847–53.

HARRIS, S., SHETH, S. A., and COHEN, M. S. (2008), 'Functional neuroimaging of belief, disbelief, and uncertainty', *Annals of Neurology*, 63: 141–7.

HARRIS, T. (1999), *Hannibal*. London: Heinemann.

HASLINGER, B., ERHARD, P., ALTENMULLER, E., SCHROEDER, U., BOECKER, H., and CEBALLOS-BAUMANN, A. O. (2005), 'Transmodal sensorimotor networks during action observation in professional pianists', *Journal of Cognitive Neuroscience*, 17: 282–93.

HASSIN, R. R., FERGUSON, M. J., SHIDLOVSKI, D., and GROSS, T. (2007), 'Subliminal exposure to national flags affects political thought and behavior', *Proceedings of the National Academy of Sciences of the United States of America*, 104: 19757–61.

HATZFELD, J. (trans. 2005), *A Time for Machetes: The Rwandan Genocide. The Killers Speak*, trans. L. Coverdale. London: Serpent's Tail.

HAUERT, C., TRAULSEN, A., BRANDT, H., NOWAK, M. A., and SIGMUND, K. (2007), 'Via freedom to coercion: the emergence of costly punishment', *Science*, 316: 1905–7.

HAUSER, M. (2006), *Moral Minds: How Nature Designed Our Universal Sense of Right and Wrong*. New York: HarperCollins.

HAUSFATER, G. and HRDY, S. B. (1984), *Infanticide: Comparative and Evolutionary Perspectives*. New York: Aldine.

HEBERLEIN, A. S. and ADOLPHS, R. (2004), 'Impaired spontaneous anthropomorphizing despite intact perception and social knowledge', *Proceedings of the National Academy of Sciences of the United States of America*, 101: 7487–91.

HEEKEREN, H. R., WARTENBURGER, I., SCHMIDT, H., PREHN, K., SCHWINTOWSKI, H. P., and VILLRINGER, A. (2005), 'Influence of bodily harm on neural correlates of semantic and moral decision-making', *NeuroImage*, 24: 887–97.

HEIDER, F. and SIMMEL, M. (1944), 'An experimental study of apparent behavior', *American Journal of Psychology*, 57: 243–59.

HENNENLOTTER, A. and SCHROEDER, U. (2006), 'Partly dissociable neural substrates for recognizing basic emotions: a critical review', *Progress in Brain Research*, 156: 443–56.

HENRICH, J. and BOYD, R. (2002), 'On modeling cognition and culture: why cultural evolution does not require replication of representations', *Journal of Cognition and Culture*, 2: 87–112.

HENSLEY, C. and TALLICHET, S. E. (2005a), 'Animal cruelty motivations: assessing demographic and situational influences', *Journal of Interpersonal Violence*, 20: 1429–43.

——— —— (2005b), 'Learning to be cruel?: exploring the onset and frequency of animal cruelty', *International Journal of Offender Therapy and Comparative Criminology*, 49: 37–47.

——— —— (2008), 'The effect of inmates' self-reported childhood and adolescent animal cruelty: motivations on the number of convictions for adult violent interpersonal crimes', *International Journal of Offender Therapy and Comparative Criminology*, 52: 175–84.

HERSH, S. M. (1972), *Cover-up: The Army's Secret Investigation of the Massacre at My Lai 4*. New York: Random House.

HERTEL, B. R. and DONAHUE, M. J. (1995), 'Parental influences on God images among children: testing Durkheim's metaphoric parallelism', *Journal for the Scientific Study of Religion*, 34: 186–99.

HILBERG, R. (1985), *The Destruction of the European Jews*, student edn. New York: Holmes & Meier.

HILLENBRAND, U. and VAN HEMMEN, J. L. (2002), 'Adaptation in the corticothalamic loop: computational prospects of tuning the senses', *Philosophical Transactions of the Royal Society Series B: Biological Sciences*, 357: 1859–67.

HIMMLER, H. (last accessed 29 May 2008), *Poznan speech*. Located at http://www.holocaust-history.org/himmler-poznan/speech-text.shtml.

HINTON, A. L., ed. (2002), *Annihilating Difference: The Anthropology of Genocide*. Berkeley: University of California Press.

—— (2004), *Why Did They Kill? Cambodia in the Shadow of Genocide*. Berkeley: University of California Press.

Hippocrates (trans. 1923), 'Epidemics I'. In *Hippocrates I*, trans. W. H. S. Jones. Cambridge, Mass.: Harvard University Press, 139–288.

HITLER, A. (trans. 1969), *Mein Kampf*, trans. R. Manheim. London: Hutchinson.

HOBBES, T., ed. (1996), *Leviathan*, ed. R. Tuck. Cambridge: Cambridge University Press.

HOCHSCHILD, A. (1999), *King Leopold's Ghost: A Story of Greed, Terror and Heroism in the Congo*. London: Macmillan.

HODSON, G. and COSTELLO, K. (2007), 'Interpersonal disgust, ideological orientations, and dehumanization as predictors of intergroup attitudes', *Psychological Science*, 18: 691–8.

HOLLEY, J. W. (1977), 'Tenure and research productivity', *Research in Higher Education*, 6: 181–92.

HOLY, T. E. (2007), 'A public confession: the retina trumpets its failed predictions', *Neuron*, 55: 831–2.

HOMER (trans. 1999), *The Iliad; The Odyssey*, ed. B. Knox and R. Fagles. London: Penguin.

VAN HONK, J. and SCHUTTER, D. J. (2007), 'Testosterone reduces conscious detection of signals serving social correction: implications for antisocial behavior', *Psychological Science*, 18: 663–7.

HORNBY, P. J. (2001), 'Central neurocircuitry associated with emesis', *American Journal of Medicine*, 11: S106–12.

HOWARD, M., ANDREOPOULOS, G. J., and SHULMAN, M. R., eds. (1994), *The Laws of War: Constraints on Warfare in the Western World*. New Haven: Yale University Press.

HUGHES, T. (1997), *Tales from Ovid*. London: Faber & Faber.

HUME, D. (1975), *Enquiries Concerning Human Understanding and Concerning the Principles of Morals*, ed. L. A. Selby-Bigge and P. H. Nidditch. Oxford: Oxford University

Press.

HYMAN, S. E., MALENKA, R. C., and NESTLER, E. J. (2006), 'Neural mechanisms of addiction: the role of reward-related learning and memory', *Annual Review of Neuroscience*, 29: 565–98.

IBARGÜENGOITIA, J. (trans. 1983), *The Dead Girls*, trans. A. Zatz. London: Chatto & Windus.

JAMES, O. (2007), *Affluenza: How to be Successful and Stay Sane*. London: Vermillion.

JANIS, I. L. (1982), *Groupthink: Psychological Studies of Policy Decisions and Fiascos*, 2nd edn. Boston: Houghton Mifflin.

JEANNEROD, M. and FRAK, V. (1999), 'Mental imaging of motor activity in humans', *Current Opinion in Neurobiology*, 9: 735–9.

JENKINS, A. C., MACRAE, C. N., and MITCHELL, J. P. (2008), 'Repetition suppression of ventromedial prefrontal activity during judgments of self and others', *Proceedings of the National Academy of Sciences of the United States of America*, 105: 4507–12.

JOHNSTONE, R. A. and BSHARY, R. (2007), 'Indirect reciprocity in asymmetric interactions: when apparent altruism facilitates profitable exploitation', *Philosophical Transactions of the Royal Society Series B: Biological Sciences*, 274: 3175–81.

JONAS, E., GRAUPMANN, V., and FREY, D. (2006), 'The influence of mood on the search for supporting versus conflicting information: dissonance reduction as a means of mood regulation?' *Personality and Social Psychology Bulletin*, 32: 3–15.

JONAS, K. J. and SASSENBERG, K. (2006), 'Knowing how to react: automatic response priming from social categories', *Journal of Personality and Social Psychology*, 90: 709–21.

JONES, D. (2007), 'Moral psychology: the depths of disgust', *Nature*, 447: 768–71.

JONES, L. M., FONTANINI, A., SADACCA, B. F., MILLER, P., and KATZ, D. B. (2007), 'Natural stimuli evoke dynamic sequences of states in sensory cortical ensembles', *Proceedings of the National Academy of Sciences of the United States of America*, 104: 18772–7.

JONES, M. and FABIAN, A., eds. (2006), *Conflict*. Cambridge: Cambridge University Press.

JOURDAIN, P., BERGERSEN, L. H., BHAUKAURALLY, K., BEZZI, P., SANTELLO, M., DOMERCQ, M., MATUTE, C., TONELLO, F., GUNDERSEN, V., and VOLTERRA, A. (2007), 'Glutamate exocytosis from astrocytes controls synaptic strength', *Nature Neuroscience*, 10: 331–9.

KASSIMERIS, G., ed. (2006), *The Barbarisation of Warfare*. London: Hurst.

KAUER, J. A. and MALENKA, R. C. (2007), 'Synaptic plasticity and addiction', *Nature Reviews Neuroscience*, 8: 844–58.

KEELEY, L. H. (1996), *War Before Civilization*. Oxford: Oxford University Press.

KEKES, J. (1996), 'Cruelty and liberalism', *Ethics*, 106: 834–44.

KELEMEN, D. (2003), 'British and American children's preferences for teleo-functional explanations of the natural world', *Cognition*, 88: 201–21.

KELLY, C., GRINBAND, J., and HIRSCH, J. (2007), 'Repeated exposure to media violence is associated with diminished response in an inhibitory frontolimbic network', *PLoS ONE*, 2, p. e1268. DOI: http://dx.doi.org/10.1371/journal.pone.0001268.

KELLY, D. J., QUINN, P. C., SLATER, A. M., LEE, K., GE, L., and PASCALIS, O. (2007), 'The other-race effect develops during infancy: evidence of perceptual narrowing', *Psychological Science*, 18: 1084–9.

KINSEY, A. C., POMEROY, W. B., MARTIN, C. E., and GEBHARD, P. H. (1953), *Sexual Behavior in the Human Female*. Philadelphia: W. B. Saunders.

KINZLER, K. D., DUPOUX, E., and SPELKE, E. S. (2007), 'The native language of social cognition', *Proceedings of the National Academy of Sciences of the United States of America*, 104: 12577–80.

KIRSCH, L. G. and BECKER, J. V. (2007), 'Emotional deficits in psychopathy and sexual sadism: implications for violent and sadistic behavior', *Clinical Psychology Review*, 27: 904–22.

KISSLER, J., HERBERT, C., PEYK, P., and JUNGHOFER, M. (2007), 'Buzzwords: early cortical responses to emotional words during reading', *Psychological Science*, 18: 475–80.

KLEE, E., DRESSEN, W., and RIESS, V., eds. (1991), *'The Good Old Days': The Holocaust as Seen by its Perpetrators and Bystanders*. New York: Konecky & Konecky.

KOENEN, K. C., MOFFITT, T. E., CASPI, A., TAYLOR, A., and PURCELL, S. (2003), 'Domestic violence is associated with environmental suppression of IQ in young children', *Development and Psychopathology*, 15: 297–311.

KOLNAI, A. ed. (2004), *On Disgust*, ed. B. Smith and C. Korsmeyer. Chicago and La Salle: Open Court.

KRAFFT-EBING, R. (trans. 1965), *Psychopathia Sexualis*, trans. F. S. Klaf, 12th edn. London: Staples Press.

KRAMER, U. M., JANSMA, H., TEMPELMANN, C., and MUNTE, T. F. (2007), 'Tit-for-tat: the neural basis of reactive aggression', *NeuroImage*, 38: 203–11.

KROLAK-SALMON, P., HENAFF, M. A., ISNARD, J., TALLON-BAUDRY, C., GUENOT, M., VIGHETTO, A., BERTRAND, O., and MAUGUIERE, F. (2003), 'An attention modulated response to disgust in human ventral anterior insula', *Annals of Neurology*, 53: 446–53.

KRUEGER, R. F., MARKON, K. E., PATRICK, C. J., BENNING, S. D., and KRAMER, M. D. (2007), 'Linking antisocial behavior, substance use, and personality: an integrative quantitative model of the adult externalizing spectrum', *Journal of Abnormal Psychology*, 116: 645–66.

KRUGLANSKI, A. W., SHAH, J. Y., PIERRO, A., and MANNETTI, L. (2002), 'When similarity breeds content: need for closure and the allure of homogeneous and self-resembling groups', *Journal of Personality and Social Psychology*, 83: 648–62.

KRUMHUBER, E., MANSTEAD, A. S. R., COSKER, D., MARSHALL, D., ROSIN, P. L., and KAPPAS, A. (2007), 'Facial dynamics as indicators of trustworthiness and cooperative behavior', *Emotion*, 7: 730–5.

KUNIECKI, M., URBANIK, A., SOBIECKA, B., KOZUB, J., and BINDER, M. (2003), 'Central control of heart rate changes during visual affective processing as revealed by fMRI', *Acta Neurobiologiae Experimentalis (Warszawa)*, 63: 39–48.

KURST-SWANGER, K. and PETCOSKY, J. L. (2003), *Violence in the Home: Multidisciplinary Perspectives*. New York: Oxford University Press.

KVERAGA, K., GHUMAN, A. S., and BAR, M. (2007), 'Top-down predictions in the cognitive brain', *Brain and Cognition*, 65: 145–68.

LACAPRA, D. (1997), 'Lanzmann's *Shoah*: here there is no why', *Critical Inquiry*, 23: 231–69.

LANDIS, B. N., LEUCHTER, I., SAN MILLAN RUIZ, D., LACROIX, J. S., and LANDIS, T. (2006), 'Transient hemiageusia in cerebrovascular lateral pontine lesions', *Journal of Neurology, Neurosurgery and Psychiatry*, 77: 680–3.

LANGER, L. L. (1999), *Preempting the Holocaust*. New Haven: Yale University Press.

LAQUEUR, W. (2004), *Voices of Terror: Manifestos, Writings and Manuals of Al Qaeda, Hamas, and Other Terrorists from Around the World and Throughout the Ages*. New York: Reed Press.

LEDOUX, J. (1998), *The Emotional Brain: The Mysterious Underpinnings of Emotional Life*. London: Weidenfeld & Nicolson.

LEE, D. K., ITTI, L., KOCH, C., and BRAUN, J. (1999), 'Attention activates winner-take-all competition among visual filters', *Nature Neuroscience*, 2: 375–81.

LEKNES, S. and TRACEY, I. (2008), 'A common neurobiology for pain and pleasure', *Nature Reviews Neuroscience*, 9: 314–20.

LESLIE, K. R., JOHNSON-FREY, S. H., and GRAFTON, S. T. (2004), 'Functional imaging of face and hand imitation: towards a motor theory of empathy', *NeuroImage*, 21: 601–7.

LEVENSON, R. W., EKMAN, P., and FRIESEN, W. V. (1990), 'Voluntary facial action generates emotion-specific autonomic nervous system activity', *Psychophysiology*, 27: 363–84.

LIBET, B., FREEMAN, A., and SUTHERLAND, K., eds. (1999), *The Volitional Brain: Towards a Neuroscience of Free Will*. Thorverton: Imprint Academic.

LIEBERMAN, D., TOOBY, J., and COSMIDES, L. (2007), 'The architecture of human kin detection', *Nature*, 445: 727–31.

LIEBERMAN, M. D., EISENBERGER, N. I., CROCKETT, M. J., TOM, S. M., PFEIFER, J. H., and WAY, B. M. (2007), 'Putting feelings into words: affect labeling disrupts amygdala activity in response to affective stimuli', *Psychological Science*, 18: 421–8.

LIFTON, R. J. (2000), *The Nazi Doctors: Medical Killing and the Psychology of Genocide*. New York: Basic Books.

LINDSAY, J. J. and ANDERSON, C. A. (2000), 'From antecedent conditions to violent actions: a general affective aggression model', *Personality and Social Psychology Bulletin*, 26: 533–47.

LINSER, K. and GOSCHKE, T. (2007), 'Unconscious modulation of the conscious experience of voluntary control', *Cognition*, 104: 459–75.

LOCHNER, L. P. (1943), *What About Germany?* London: Hodder & Stoughton.

LOZA, W. (2007), 'The psychology of extremism and terrorism: a Middle-Eastern perspective', *Aggression and Violent Behavior*, 12: 141–55.

LUO, Q., HOLROYD, T., JONES, M., HENDLER, T., and BLAIR, J. (2007), 'Neural dynamics for facial threat processing as revealed by gamma band synchronization using MEG', *NeuroImage*, 34: 839–47.

LYNCH, G., REX, C. S., and GALL, C. M. (2007), 'LTP consolidation: substrates, explanatory power, and functional significance', *Neuropharmacology*, 52: 12–23.

MCARTHUR, B. (2006), *Surviving the Sword: Prisoners of the Japanese, 1942–45*. London: Abacus.

MCCALL, G. S. and SHIELDS, N. (2008), 'Examining the evidence from small-scale societies and early prehistory and implications for modern theories of aggression and violence', *Aggression and Violent Behavior*, 13: 1–9.

MACHIAVELLI, N. (trans. 1961), *The Prince*, trans. G. Bull. Harmondsworth: Penguin.

MACKAY, C. (ed. 1973), *Selections from 'Extraordinary Popular Delusions and the Madness of Crowds'*. London: Unwin.

MCKELVIE, S. J. and COLEY, J. (1993), 'Effects of crime seriousness and offender facial attractiveness on recommended treatment', *Social Behavior and Personality*, 21: 265–77.

MAJDANDZIC, J., GROL, M. J., VAN SCHIE, H. T., VERHAGEN, L., TONI, I., and BEKKERING, H. (2007), 'The role of immediate and final goals in action planning: an fMRI study', *NeuroImage*, 37: 589–98.

MALCOLM, N. (1996), *Bosnia: A Short History*. London: Papermac.

—— (1998), *Kosovo: A Short History*. London: Macmillan.

MALLE, B. F. (2006), 'The actor–observer asymmetry in attribution: a (surprising) meta-analysis', *Psychological Bulletin*, 132: 895–919.

MANER, J. K., KENRICK, D. T., BECKER, D. V., ROBERTSON, T. E., HOFER, B., NEUBERG, S. L., DELTON, A. W., BUTNER, J., and SCHALLER, M. (2005), 'Functional projection: how fundamental social motives can bias interpersonal perception', *Journal of Personality and Social Psychology*, 88: 63–78.

MARSHALL, W. L. and KENNEDY, P. (2003), 'Sexual sadism in sexual offenders: an elusive diagnosis', *Aggression and Violent Behavior*, 8: 1–22.

MARTENS, A., KOSLOFF, S., GREENBERG, J., LANDAU, M. J., and SCHMADER, T. (2007), 'Killing begets killing: evidence from a bug-killing paradigm that initial killing fuels subsequent killing', *Personality and Social Psychology Bulletin*, 33: 1251–64.

MARUYA, K., YANG, E., and BLAKE, R. (2007), 'Voluntary action influences visual competition', *Psychological Science*, 18: 1090–8.

MASSEY, P. V. and BASHIR, Z. I. (2007), 'Long-term depression: multiple forms and implications for brain function', *Trends in Neurosciences*, 30: 176–84.

MAXWELL, J. S. and DAVIDSON, R. J. (2007), 'Emotion as motion: asymmetries in approach and avoidant actions', *Psychological Science*, 18: 1113–19.

MAY, A. (2007), 'Neuroimaging: visualising the brain in pain', *Neurological Sciences*, 28: S101–7.

MELSON, R. (1992), *Revolution and Genocide: On the Origins of the Armenian Genocide and the Holocaust*. Chicago: University of Chicago Press.

MERTUS, J. A. (1999), *Kosovo: How Myths and Truths Started a War*. Berkeley: University of California Press

MERZ-PEREZ, L. and HEIDE, K. M. (2003), *Animal Cruelty: Pathway to Violence against People*. Lanham, Md.: Altamira Press.

METCALFE, J. and GREENE, M. J. (2007), 'Metacognition of agency', *Journal of Experimental Psychology: General*, 136: 184–99.

MIKHAIL, J. (2007), 'Universal moral grammar: theory, evidence and the future', *Trends in Cognitive Sciences*, 11: 143–52.

MIKKELSON, B. and MIKKELSON, D. P. (last accessed 29 May 2008), *The Twinkie Defense*. Located at http://www.snopes.com/legal/twinkie.asp.

MILGRAM, S. (1963), 'Behavioral study of obedience', *Journal of Abnormal and Social Psychology*, 67: 371–8.

—— (1997), *Obedience to Authority*. London: Pinter & Martin.

MILLER, J. (1990), 'Carnivals of atrocity: Foucault, Nietzsche, cruelty', *Political Theory*, 18: 470–91.

MILLER, S. B. (2004), *Disgust: The Gatekeeper Emotion*. Hillsdale, NJ: Analytic Press.

MILLER, W. I. (1997), *The Anatomy of Disgust*. Cambridge, Mass.: Harvard University Press.

MIYATA, M. (2007), 'Distinct properties of corticothalamic and primary sensory synapses to thalamic neurons', *Neuroscience Research*, 59: 377–82.

MONK, R. (1990), *Ludwig Wittgenstein: The Duty of Genius*. London: Cape.

MOORE, R. I. (1987), *The Formation of a Persecuting Society: Power and Deviance in Western Europe, 950–1250*. Oxford: Blackwell.

MORGENTHAU, H. (ed. 2000), *Ambassador Morgenthau's Story*. Reading: Taderon Press, by arrangement with the Gomidas Institute.

MORRISON, I. and DOWNING, P. E. (2007), 'Organization of felt and seen pain responses in anterior cingulate cortex', *NeuroImage*, 37: 642–51.

MURIS, P., MAYER, B., HUIJDING, J., and KONINGS, T. (2008), 'A dirty animal is a scary animal! Effects of disgust-related information on fear beliefs in children', *Behaviour Research and Therapy*, 46: 263–9.

NAGEL, T. (1972), 'War and massacre', *Philosophy and Public Affairs*, 1: 123–44.

NAGENGAST, C. (2002), 'Inoculation of evil in the U.S.–Mexican border region: reflections on the genocidal potential of symbolic violence', in A. L. Hinton (ed.), *Annihilating Difference: The Anthropology of Genocide*. Berkeley: University of California Press, 325–47.

NAIDICH, T. P., KANG, E., FATTERPEKAR, G. M., DELMAN, B. N., GULTEKIN, S. H., WOLFE, D., ORTIZ, O., YOUSRY, I., WEISMANN, M., and YOUSRY, T. A. (2004), 'The insula: anatomic study and MR imaging display at 1.5 T', *American Journal of Neuroradiology*, 25: 222–32.

NAVARRETE, C. D. and FESSLER, D. M. T. (2006), 'Disease avoidance and ethnocentrism: the effects of disease vulnerability and disgust sensitivity on intergroup attitudes', *Evolution and Human Behavior*, 27: 270–82.

NELL, V. (2006), 'Cruelty's rewards: the gratifications of perpetrators and spectators', *Behavioral and Brain Sciences*, 29: 211–24; discussion, pp. 224–57.

NEW, J., COSMIDES, L., and TOOBY, J. (2007), 'Category-specific attention for animals reflects ancestral priorities, not expertise', *Proceedings of the National Academy of Sciences of the United States of America*, 104: 16598–603.

NEWMAN, L. S. and ERBER, R., eds. (2002), *Understanding Genocide: The Social Psychology of the Holocaust*. Oxford: Oxford University Press.

NG-MAK, D. S., SALZINGER, S., FELDMAN, R. S., and STUEVE, C. A. (2004), 'Pathologic adaptation to community violence among inner-city youth', *American Journal of Orthopsychiatry*, 74: 196–208.

NICHOLAS, S., KERSHAW, C., and WALKER, A., eds. (2007), *Crime in England and Wales 2006/07*. London: Research, Development and Statistics Directorate.

NIETZSCHE, F. W. (trans. 1968), *The Anti-Christ*. In *Twilight of the Idols and The Anti-Christ*, trans. R. J. Hollingdale. Harmondsworth: Penguin.

——(trans. 1973), *Beyond Good and Evil*, trans. R. J. Hollingdale. Harmondsworth: Penguin.

Nisbett, R. E. and Cohen, D. (1996), *Culture of Honor: The Psychology of Violence in the South*. Boulder, Colo.: Westview.

Nordgren, L. F., van der Pligt, J., and van Harreveld, F. (2007), 'Evaluating Eve: visceral states influence the evaluation of impulsive behavior', *Journal of Personality and Social Psychology*, 93: 75–84.

Nuland, S. B. (1994), *How We Die*. London: Chatto & Windus.

Nussbaum, S., Trope, Y., and Liberman, N. (2003), 'Creeping dispositionism: the temporal dynamics of behaviour prediction', *Journal of Personality and Social Psychology*, 84: 485–97.

Offer, A. (2006), *The Challenge of Affluence: Self-control and Well-being in the United States and Britain since 1950*. Oxford: Oxford University Press.

Ohman, A., Carlsson, K., Lundqvist, D., and Ingvar, M. (2007), 'On the unconscious subcortical origin of human fear', *Physiology and Behavior*, 92: 180–5.

Overy, R. (2001), *Interrogations: The Nazi Elite in Allied Hands, 1945*. London: Allen Lane.

Papafragou, A., Cassidy, K., and Gleitman, L. (2007), 'When we think about thinking: the acquisition of belief verbs', *Cognition*, 105: 125–65.

Park, J. H., Faulkner, J., and Schaller, M. (2003), 'Evolved disease-avoidance processes and contemporary anti-social behavior: prejudicial attitudes and avoidance of people with physical disabilities', *Journal of Nonverbal Behavior*, 27: 65–87.

Parker, R. (1983), *Miasma*. Oxford: Clarendon Press.

Pavlov, I. P. (trans. 1941), *Lectures on Conditioned Reflexes*. Volume 2. *Conditioned Reflexes and Psychiatry*, trans. W. H. Gantt. London: Lawrence & Wishart.

Peper, M. (2006), 'Imaging emotional brain functions: conceptual and methodological issues', *Journal of Physiology (Paris)*, 99: 293–307.

Pexman, P. M., Hargreaves, I. S., Edwards, J. D., Henry, L. C., and Goodyear, B. G. (2007), 'The neural consequences of semantic richness: when more comes to mind, less activation is observed', *Psychological Science*, 18: 401–6.

Pfurtscheller, G., Neuper, C., Ramoser, H., and Muller-Gerking, J. (1999), 'Visually guided motor imagery activates sensorimotor areas in humans', *Neuroscience Letters*, 269: 153–6.

Phillips, M. L., Senior, C., Fahy, T., and David, A. S. (1998), 'Disgust—the forgotten emotion of psychiatry', *British Journal of Psychiatry*, 172: 373–5.

Phillips, M. L., Williams, L. M., Heining, M., Herba, C. M., Russell, T., Andrew, C., Bullmore, E. T., Brammer, M. J., Williams, S. C., Morgan, M., Young, A. W., and Gray, J. A. (2004), 'Differential neural responses to overt and covert presentations of facial expressions of fear and disgust', *NeuroImage*, 21: 1484–96.

PIERRO, A., MANNETTI, L., DE GRADA, E., LIVI, S., and KRUGLANSKI, A. W. (2002), 'Autocracy bias in informal groups under need for closure', *Personality and Social Psychology Bulletin*, 29: 405–17.

PINCUS, J. H. (2001), *Base Instincts: What Makes Killers Kill?* New York: W. W. Norton.

PLAKS, J. E., GRANT, H., and DWECK, C. S. (2005), 'Violations of implicit theories and the sense of prediction and control: implications for motivated person perception', *Journal of Personality and Social Psychology*, 88: 245–62.

PLATEK, S. M., KRILL, A. L., and KEMP, S. M. (2008), 'The neural basis of facial resemblance', *Neuroscience Letters*, 437: 76–81.

PORTER, S., WOODWORTH, M., EARLE, J., DRUGGE, J., and BOER, D. (2003), 'Characteristics of sexual homicides committed by psychopathic and nonpsychopathic offenders', *Law and Human Behavior*, 27: 459–70.

POWER, S. (2003), *'A Problem from Hell': America and the Age of Genocide*. London: Flamingo.

PRESTON, R. (1994), *The Hot Zone*. London: Doubleday.

PRUNIER, G. (1995), *The Rwanda Crisis*. New York: Columbia University Press.

PUGH, M. (2006), *'Hurrah for the Blackshirts!' Fascists and Fascism in Britain Between the Wars*. London: Pimlico.

RAAFLAUB, K. A., OBER, J., and WALLACE, R. W., eds. (2007), *Origins of Democracy in Ancient Greece*. Berkeley: University of California Press.

RAYMOND, C. R. (2007), 'LTP forms 1, 2 and 3: different mechanisms for the "long" in long-term potentiation', *Trends in Neurosciences*, 30: 167–75.

REES, L. (2005), *Auschwitz: The Nazis and the Final Solution*. London: BBC Books.

REJALI, D. (2008), *Torture and Democracy*. Princeton: Princeton University Press.

RESTORATIVE JUSTICE CONSORTIUM (last accessed 29 May 2008), *Restorative Justice Consortium*. Located at http://www.restorativejustice.org.uk.

RHODES, R. (1988), *The Making of the Atomic Bomb*. Harmondsworth: Penguin.

—— (2003), *Masters of Death: The SS-Einsatzgruppen and the Invention of the Holocaust*. New York: Vintage.

RICHARDSON, D. S. and HAMMOCK, G. S. (2007), 'Social context of human aggression: are we paying too much attention to gender?', *Aggression and Violent Behavior*, 12: 417–26.

RITZ, T., THONS, M., FAHRENKRUG, S., and DAHME, B. (2005), 'Airways, respiration, and respiratory sinus arrhythmia during picture viewing', *Psychophysiology*, 42: 568–78.

RIZZOLATTI, G. (2005), 'The mirror neuron system and its function in humans', *Anatomy and Embryology (Berlin)*, 210: 419–21.

Robins, R. S. and Post, J. M. (1997), *Political Paranoia: The Psychopolitics of Hatred*. New Haven: Yale University Press.

Rogers, D. S. and Ehrlich, P. R. (2008), 'Natural selection and cultural rates of change', *Proceedings of the National Academy of Sciences of the United States of America*, 105: 3416–20.

de Roos, S. A., Iedema, J., and Miedema, S. (2004), 'Influence of maternal denomination, God concepts, and child-rearing practices on young children's God concepts', *Journal for the Scientific Study of Religion*, 43: 519–35.

Rosenbaum, A. S., ed. (2000), *Is the Holocaust Unique? Perspectives on Comparative Genocide*, 2nd edn. Boulder, Colo.: Westview Press.

Rousseau, J. J. (trans. 1973), *The Social Contract and Discourses*, trans. G. D. H. Cole. London: Dent.

Rozin, P. and Fallon, A. E. (1987), 'A perspective on disgust', *Psychological Review*, 94: 23–41.

Rozin, P., Millman, L., and Nemeroff, C. (1986), 'Operation of the laws of sympathetic magic in disgust and other domains', *Journal of Personality and Social Psychology*, 50: 703–12.

Rummel, R. J. (1994), *Death by Government*. New Brunswick, NJ: Transaction.

Rushdie, S. (1991), *Haroun and the Sea of Stories*. London: Granta.

Russell, B. and Whitehead, A. N. (1910–13), *Principia Mathematica*. Cambridge: Cambridge University Press.

Ruthven, M. (2004), *Fundamentalism: The Search for Meaning*. Oxford: Oxford University Press.

Ryle, G. (1971a), 'Thinking and reflecting'. In *Collected Papers*. Volume 2. *Collected Essays 1929–1968*. London: Hutchinson, 465–79.

—— (1971b), 'The thinking of thoughts: what is "Le Penseur" doing?' In *Collected Papers*. Volume 2. *Collected Essays 1929–1968*. London: Hutchinson, 480–96.

Sade, D. A. F., Marquis de (trans. 1989), *120 Days of Sodom*, trans. A. Wainhouse and R. Seaver. London: Arrow.

—— (trans. 1991), *Justine, or Good Conduct Well Chastised*. In *Justine, Philosophy in the Bedroom and other Writings*, trans. A. Wainhouse and R. Seaver. London: Arrow, 447–743.

Saha, S. (2005), 'Role of the central nucleus of the amygdala in the control of blood pressure: descending pathways to medullary cardiovascular nuclei', *Clinical and Experimental Pharmacology and Physiology (Victoria)*, 32: 450–6.

Saito, R., Takano, Y., and Kamiya, H. (2003), 'Roles of Substance P and NK1 receptor in the brainstem in the development of emesis', *Journal of Pharmacological Sciences*, 91: 87–94.

SALMINEN, M. and RAVAJA, N. (2008), 'Increased oscillatory theta activation evoked by violent digital game events', *Neuroscience Letters*, 435: 69–72.

SAMPSON, A. (1993), Acts of Abuse: Sex Offenders and the Criminal Justice System. London: Routledge.

SARLO, M., BUODO, G., POLI, S., and PALOMBA, D. (2005), 'Changes in EEG alpha power to different disgust elicitors: the specificity of mutilations', *Neuroscience Letters*, 382: 291–6.

SCHACHTER, S. and SINGER, J. E. (1962), 'Cognitive, social, and physiological determinants of emotional state', *Psychological Review*, 69: 379–99.

SCHERER, K. R. and WALLBOTT, H. G. (1994), 'Evidence for universality and cultural variation of differential emotion response patterning', *Journal of Personality and Social Psychology*, 66: 310–28.

SCHULTZ, W., DAYAN, P., and MONTAGUE, P. R. (1997), 'A neural substrate of prediction and reward', *Science*, 275: 1593–9.

SEMELIN, J. (trans. 2007), *Purify and Destroy*, trans. C. Schoch. London: Hurst.

SENECA (trans. 2007), 'On Mercy'. In *Dialogues and Essays*, trans. J. Davie. Oxford: Oxford University Press, 188–218.

SEQUEIRA, H., VILTART, O., BA-M'HAMED, S., AND POULAIN, P. (2000), 'Cortical control of somato-cardiovascular integration: neuroanatomical studies', *Brain Research Bulletin*, 53: 87–93.

SEWARDS, T.V. (2004), 'Dual separate pathways for sensory and hedonic aspects of taste', *Brain Research Bulletin*, 62: 271–83.

SHAKESPEARE, W. (ed. 1997), *King Lear*, ed. R. A. Foakes. London: Arden Shakespeare.

SHAMAY-TSOORY, S. G. and AHARON-PERETZ, J. (2007), 'Dissociable prefrontal networks for cognitive and affective theory of mind: a lesion study', *Neuropsychologia*, 45: 3054–67.

SHAROT, T., RICCARDI, A. M., RAIO, C. M., and PHELPS, E. A. (2007), 'Neural mechanisms mediating optimism bias', *Nature*, 450: 102–5.

SHEMA, R., SACKTOR, T. C., and DUDAI, Y. (2007), 'Rapid erasure of long-term memory associations in the cortex by an inhibitor of PKM-zeta', *Science*, 317: 951–3.

SHERMAN, L. W. and STRANG, H. (last accessed 29 May 2008), *Restorative Justice: The Evidence*. London: Smith Institute. Located at http://www.esmeefairbairn.org.uk/docs/RJ_full_report.pdf.

SHKLAR, J. N. (1984), *Ordinary Vices*. Cambridge, Mass.: Belknap.

SIMION, F., REGOLIN, L., and BULF, H. (2008), 'A predisposition for biological motion in the newborn baby', *Proceedings of the National Academy of Sciences of the United States of America*, 105: 809–13.

SINGER, P. (1971), 'Famine, affluence, and morality', *Philosophy and Public Affairs*, 1: 229–43.

SINGER, T., KIEBEL, S. J., WINSTON, J. S., DOLAN, R. J., and FRITH, C. D. (2004), 'Brain responses to the acquired moral status of faces', *Neuron*, 41: 653–62.

—— SEYMOUR, B., O'DOHERTY, J. P., STEPHAN, K. E., DOLAN, R. J., AND FRITH, C. D. (2006), 'Empathic neural responses are modulated by the perceived fairness of others', *Nature*, 439: 466–9.

SLOTE, M. (1990), 'Ethics without free will', *Social Theory and Practice*, 16: 369–83.

SORABJI, R. and RODIN, D. (2006), *The Ethics of War: Shared Problems in Different Traditions*. Aldershot: Ashgate.

SPARROW, B. and WEGNER, D. M. (2006), 'Unpriming: the deactivation of thoughts through expression', *Journal of Personality and Social Psychology*, 91: 1009–19.

STANTON, G. H. (last accessed 29 May 2008), *The 8 stages of genocide*. Washington, DC: Genocide Watch. Located at http://www.genocidewatch.org/8stages.htm.

STARK, R., WALTER, B., SCHEINLE, A., and VAITL, D. (2005), 'Psychophysiological correlates of disgust and disgust sensitivity', *Journal of Psychophysiology*, 19: 50 60.

STAUB, E. (2003), *The Psychology of Good and Evil: Why Children, Adults, and Groups Help and Harm Others*. New York: Cambridge University Press.

STERN, E. R., WAGER, T. D., EGNER, T., HIRSCH, J., and MANGELS, J. A. (2007), 'Preparatory neural activity predicts performance on a conflict task', *Brain Research*, 1176: 92–102.

STERNBERG, R. J., ed. (2005), *The Psychology of Hate*. Washington, DC: American Psychological Association.

STERZER, P., RUSS, M. O., PREIBISCH, C., and KLEINSCHMIDT, A. (2002), 'Neural correlates of spontaneous direction reversals in ambiguous apparent visual motion', *NeuroImage*, 15: 908–16.

—— STADLER, C., POUSTKA, F., and KLEINSCHMIDT, A. (2007), 'A structural neural deficit in adolescents with conduct disorder and its association with lack of empathy', *NeuroImage*, 37: 335–42.

STETS, J. (1995), 'Job autonomy and control over one's spouse: a compensatory process', *Journal of Health and Social Behavior*, 36: 244–58.

STEVENSON, R. J. and REPACHOLI, B. M. (2005), 'Does the source of an interpersonal odour affect disgust? A disease risk model and its alternatives', *European Journal of Social Psychology*, 35: 375–401.

STONE, A. and VALENTINE, T. (2005), 'Orientation of attention to nonconsciously recognised famous faces', *Cognition and Emotion*, 19: 537–58.

SUMNER, W. G. (1907), *Folkways: A Study of the Sociological Importance of Usages, Manners, Customs, Mores and Morals.* Boston: Ginn.

SURIAN, L., CALDI, S., and SPERBER, D. (2007), 'Attribution of beliefs by 13-month-old infants', *Psychological Science*, 18: 580–6.

TAJFEL, H., FLAMENT, C., BILLIG, M. G., and BUNDY, R. P. (1971), 'Social categorization and intergroup behaviour', *European Journal of Social Psychology*, 1: 149–78.

TAYLOR, K. E. (2001), 'Applying continuous modelling to consciousness', *Journal of Consciousness Studies*, 8: 45–60.

TAYLOR, K. (2004), *Brainwashing: The Science of Thought Control.* Oxford: Oxford University Press.

——— (2006), 'On brainwashing'. In G. Kassimeris (ed.), *The Barbarisation of Warfare.* London: Hurst, 238–53.

——— (2007), 'Disgust is a factor in extreme prejudice', *British Journal of Social Psychology*, 43: 597–617.

TAYLOR, S. E. and GOLLWITZER, P. M. (1995), 'Effects of mindset on positive illusions', *Journal of Personality and Social Psychology*, 69: 213–26.

THAGARD, P., KROON, F., NERB, J., SAHDRA, B., SHELLEY, C., and WAGAR, B. (2006), *Hot Thought: Mechanisms and Applications of Emotional Cognition.* Cambridge, Mass.: MIT Press.

THIAGARAJAN, T. C., LINDSKOG, M., MALGAROLI, A., and TSIEN, R. W. (2007), 'LTP and adaptation to inactivity: overlapping mechanisms and implications for metaplasticity', *Neuropharmacology*, 52: 156–75.

TILLY, C. (2003), *The Politics of Collective Violence.* New York: Cambridge University Press.

TORMALA, Z. L. and PETTY, R. E. (2002), 'What doesn't kill me makes me stronger: the effects of resisting persuasion on attitude certainty', *Journal of Personality and Social Psychology*, 83: 1298–1313.

TRAFIMOW, D., BROMGARD, I. K., FINLAY, K. A., and KETELAAR, T. (2005), 'The role of affect in determining the attributional weight of immoral behaviors', *Personality and Social Psychology Bulletin*, 31: 935–48.

TRAVAGLI, R. A. and ROGERS, R. C. (2001), 'Receptors and transmission in the brain–gut axis: potential for novel therapies: V. Fast and slow extrinsic modulation of dorsal vagal complex circuits', *American Journal of Physiolog—Gastrointestinal and Liver Physiology*, 281: 595–601.

——— HERMANN, G. E., BROWNING, K. N., and ROGERS, R. C. (2003), 'Musings on the wanderer: what's new in our understanding of vago-vagal reflexes?: III. Activity-dependent plasticity in vago-vagal reflexes controlling the stomach', *American Journal of Physiology—Gastrointestinal and Liver Physiology*, 284: 180–7.

TSE, D., LANGSTON, R. F., KAKEYAMA, M., BETHUS, I., SPOONER, P. A., WOOD, E. R., WITTER, M. P., and MORRIS, R. G. M. (2007), 'Schemas and memory consolidation', *Science*, 316: 76–82.

ULLOA, E. R. and PINEDA, J. A. (2007), 'Recognition of point-light biological motion: mu rhythms and mirror neuron activity', *Behavioural Brain Research*, 183: 188–94.

US Department of Justice (last accessed 29 May 2008), *US Bureau of Justice Statistics: Homicide Trends in the U.S. Trends by Gender*. Located at http://www.ojp.usdoj.gov/bjs/homicide/gender.htm.

VAISH, A., GROSSMANN, T., and WOODWARD, A. (2008), 'Not all emotions are created equal: the negativity bias in social-emotional development', *Psychological Bulletin*, 134: 383–403.

VALDESOLO, P. and DESTENO, D. (2007), 'Moral hypocrisy: social groups and the flexibility of virtue', *Psychological Science*, 18: 689–90.

VALENTINO, B. A. (2004), *Final Solutions: Mass Killing and Genocide in the Twentieth Century*. Ithaca, NY: Cornell University Press.

VALERIANI, M., BETTI, V., LE PERA, D., DE ARMAS, L., MILIUCCI, R., RESTUCCIA, D., AVENANTI, A., and AGLIOTI, S. M. (2008), 'Seeing the pain of others while being in pain: a laser-evoked potentials study', *NeuroImage*, 40: 1419–28.

VALLIN, J., MESLÉ, F., ADAMETS, S., AND PYROZHKOV, S. (2002), 'A new estimate of Ukrainian population losses during the crises of the 1930s and 1940s', *Population Studies*, 56: 249–64.

VENTURA, S. J., MOSHER, W. D., CURTIN, S. C., ABMA, J. C., and HENSHAW, S. (1999), 'Highlights of trends in pregnancies and pregnancy rates by outcome: estimates for the United States, 1976–96', *National Vital Statistics Reports*, 47: 1–9. Located at http://www.ncbi.nlm.nih.gov/entrez/query.fcgi?cmd=Retrieve&db=PubMed&dopt=Citation&list_uids=10635682.

DE VIGNEMONT, F. and SINGER, T. (2006), 'The empathic brain: how, when and why?' *Trends in Cognitive Sciences*, 10: 435–41.

VIZI, E. S. and MIKE, A. (2006), 'Nonsynaptic receptors for GABA and glutamate', *Current Topics in Medicinal Chemistry*, 6: 941–8.

DE WAAL, F. B. M. (1996a), *Good Natured: The Origins of Right and Wrong in Humans and Other Animals*. Cambridge, Mass.: Harvard University Press.

—— (1996b), *Our Inner Ape: The Best and Worst of Human Nature*. London: Granta.

WALDMANN, M. R. and DIETERICH, J. H. (2007), 'Throwing a bomb on a person versus throwing a person on a bomb: intervention myopia in moral intuitions', *Psychological Science*, 18: 247–53.

WALLACE, B., CESARINI, D., LICHTENSTEIN, P., and JOHANNESSON, M. (2007), 'Heritability of ultimatum game responder behavior', *Proceedings of the National Academy of Sciences of the United States of America*, 104: 15631–4.

WALLER, J. (2002), *Becoming Evil: How Ordinary People Commit Genocide and Mass Killing*. New York: Oxford University Press.

WEGNER, D. M. (2002), *The Illusion of Conscious Will*. London: MIT Press.

WEITZ, E. D. (2003), *A Century of Genocide: Utopias of Race and Nation*. Princeton: Princeton University Press.

WHEATLEY, T. and HAIDT, J. (2005), 'Hypnotic disgust makes moral judgments more severe', *Psychological Science*, 16: 780–4.

WHITE, P. A. (2006), 'The causal asymmetry', *Psychological Review*, 113: 132–47.

WHITLOCK, J., ECKENRODE, J., and SILVERMAN, D. (2006), 'Self-injurious behaviors in a college population', *Pediatrics*, 117: 1939–48.

WICKER, B., KEYSERS, C., PLAILLY, J., ROYET, J.P., GALLESE, V., and RIZZOLATTI, G. (2003), 'Both of us disgusted in my insula: the common neural basis of seeing and feeling disgust', *Neuron*, 40: 655–64.

WILKINSON, R. (2005), *The Impact of Inequality: How to Make Sick Societies Healthier*. New York: The New Press.

WILLIAMS, E. (1967), *Beyond Belief: A Chronicle of Murder and Its Detection*. London: Hamish Hamilton.

WILSON, T. (2002), *Strangers to Ourselves: Discovering the Adaptive Unconscious*. Cambridge, Mass.: Harvard University Press.

WITTGENSTEIN, L. (trans. 1974), *Philosophical Investigations*, trans. G. E. M. Anscombe, 3rd edn. Oxford: Blackwell.

——(trans. 2001), *Tractatus Logico-Philosophicus*; Introduction by Bertrand Russell, trans. D. Pears and B. McGuinness. London: Routledge.

WOLFE, R. A., BEYER, J. M., BLACKBURN, R. T., GREENHALGH, L., NAYYAR, P. R., and SETH, A. (1996), 'Rethinking the tenure process: the influences and consequences of power and culture', *Journal of Management Inquiry*, 5: 221–36.

WOLPERT, D. M., MIALL, R. C., and KAWATO, M. (1998), 'Internal models in the cerebellum', *Trends in Cognitive Sciences*, 2: 338–47.

WOODWARD, T. S., BUCHY, L., MORITZ, S., and LIOTTI, M. (2007), 'A bias against disconfirmatory evidence is associated with delusion proneness in a nonclinical sample', *Schizophrenia Bulletin*, 33: 1023–8.

WU, F. and HUBERMAN, B. A. (2007), 'Novelty and collective attention', *Proceedings of the National Academy of Sciences of the United States of America*, 104: 17599–601.

WYER, R. S., ed. (1997), *The Automaticity of Everyday Life. Advances in Social Cognition.* Volume 10. Mahwah, NJ: Lawrence Erlbaum.

WYNDHAM, J. (1960), *The Midwich Cuckoos*. Harmondsworth: Penguin.

YU, C. and SMITH, L. B. (2007), 'Rapid word learning under uncertainty via cross-situational statistics', *Psychological Science*, 18: 414–20.

ZHONG, C. B. and LILJENQUIST, K. (2006), 'Washing away your sins: threatened morality and physical cleansing', *Science*, 313: 1451–2.

ZIMBARDO, P. (2007), *The Lucifer Effect: Understanding How Good People Turn Evil.* London: Rider.

Index